Hawker's Secret Projects

Hawker's Secret Projects

Cold War Aircraft That Never Flew

Christoper Budgen

AIR WORLD

First published in Great Britain in 2023 by
Air World
An imprint of Pen & Sword Books Ltd
Yorkshire – Philadelphia

Copyright © Christopher Budgen 2023

ISBN 978 1 39904 790 6

The right of Christoper Budgen to be identified as Author of this work has been asserted by him in accordance with the Copyright, Designs and Patents Act 1988.

A CIP catalogue record for this book is
available from the British Library.

All rights reserved. No part of this book may be reproduced or transmitted in any form or by any means, electronic or mechanical including photocopying, recording or by any information storage and retrieval system, without permission from the Publisher in writing.

Typeset by Mac Style
Printed in the UK by CPI Group (UK) Ltd, Croydon, CR0 4YY.

Pen & Sword Books Limited incorporates the imprints of After the Battle, Atlas, Archaeology, Aviation, Discovery, Family History, Fiction, History, Maritime, Military, Military Classics, Politics, Select, Transport, True Crime, Air World, Frontline Publishing, Leo Cooper, Remember When, Seaforth Publishing, The Praetorian Press, Wharncliffe Local History, Wharncliffe Transport, Wharncliffe True Crime and White Owl.

For a complete list of Pen & Sword titles please contact

PEN & SWORD BOOKS LIMITED
47 Church Street, Barnsley, South Yorkshire, S70 2AS, England
E-mail: enquiries@pen-and-sword.co.uk
Website: www.pen-and-sword.co.uk
or
PEN AND SWORD BOOKS
1950 Lawrence Rd, Havertown, PA 19083, USA
E-mail: uspen-and-sword@casematepublishers.com
Website: www.penandswordbooks.com

Contents

Acknowledgements		vi
Abbreviations and Glossary		vii
Introduction		x
Chapter 1	Hawker Aircraft, Origins and Development	1
Chapter 2	The Hawker Design Office	23
Chapter 3	Hunter Developments	47
Chapter 4	The Search for a Supersonic Fighter	73
Chapter 5	Going Up: V/STOL Studies	100
Chapter 6	SABA: Small Agile Battlefield Aircraft	129
Chapter 7	ASTOVL: The Supersonic Ambition	151
Chapter 8	UFA: Unmanned Fighter Aircraft	172
Chapter 9	Conclusion	191
Appendix: Hawker, Hawker Siddeley Aviation & BAe Kingston Project Numbers 1940–1988		202
Notes		212
Index		216

Acknowledgements

As with any work of this scope, a number of individuals and organisations have been kind enough to offer their assistance and inevitably there may be those whose names have slipped through the net.

In particular, I would like to extend my grateful thanks to Chris Farara at Brooklands Museum for his advice and proofreading assistance. I would also like to thank Dr Michael Pryce for his thoughts on some of the material and Andy Jones, former HSA Chief Test Pilot, for material relating to SABA. The late Roy Braybrook offered help and encouragement with background to the Project Office before his untimely death; Peter Amos supplied copious background information from his treasure house of knowledge; Brian Buss regaled me with his experiences at Hawker, Kingston while David Hassard was most helpful in providing information on Charles Plantin and Kingston's history.

I would also like to thank Andrew Lewis and Brooklands Museum, the staff at The National Archives for access to documents and Sue for her undoubted proofing skills.

All images used are courtesy of BAE Systems and Brooklands Museum unless otherwise noted.

Glossary

A&AEE	Aeroplane and Armament Experimental Establishment, Boscombe Down
AEW	Airborne Early Warning
AI	Airborne Interception
AM	Air Marshal
AMRAAM	Advanced Medium-Range Air-to-Air Missile
APG	Advanced Projects Group
ARA	Aircraft Research Association
ASRAAM	Advanced Short-Range Air-to-Air Missile
ASSF	Air Superiority Strike Fighter
ASTOVL	Advanced Short Take-Off and Vertical Landing
AVM	Air Vice-Marshal
AWACS	Airborne Warning and Control System
BAC	British Aircraft Corporation
BAe	British Aerospace
BVR	Beyond Visual Range
CAP	Combat Air Patrol
CAS	Chief of the Air Staff
CD	Chief Designer
CG	Centre of Gravity
CH	Chain Home (radar)
Clang Box	Diverter valve fitted within the jet pipe
COIN	Counter Insurgency
CRT	Cathode Ray Tube
CS/A	Comptroller of Supplies/Air
CTOVL	Conventional Take-Off and Vertical Landing
DCAS/OR	Deputy Chief of the Air Staff – Operational Requirements
DDOR	Deputy Director of Operational Requirements
DGGW	Directorate General of Guided Weapons
DGSR/A	Director General of Scientific Research – Air

DGTD	Director General of Technical Development
DH	de Havilland
DMARD	Department of Military Aircraft Research and Development
DO	Drawing Office
DoD	Department of Defense (USA)
DOR/C	Director of Operational Requirements – Contracts
ECU	Engine Change Unit
EDO	Experimental Drawing Office
EFA	European Fighter Aircraft
ESM	Electronic Support Measures
EWO	Essential Work Order
FEBA	Forward Edge of Battle Area
FGA	Fighter, Ground Attack
FLIR	Forward Looking Infra-Red
GA	General Arrangement
HGHE	HG Hawker Engineering
HOJ	Home on Jam
HSA	Hawker Siddeley Aviation
ILS	Instrument Landing System
IRST	Infra-Red Search and Track
ITP	Instruction To Proceed
LERX	Leading Edge Root Extension
LWR	Laser Warning Receiver
MAD	Military Aircraft Division
MDC	McDonnell Douglas Corporation
MoA	Ministry of Aviation
MoD	Ministry of Defence
MoS	Ministry of Supply
MRCA	Multi-Role Combat Aircraft
NATO	North Atlantic Treaty Organisation
NBMR	NATO Basic Military Requirement
NGTE	National Gas Turbine Establishment, Pyestock
NPL	National Physical Laboratory
OR	Operational Requirement
PCB	Plenum Chamber Burning
PDO	Production Design Office
PDSR	Department of Scientific Research

PE	Procurement Executive
PO	Project Office
RAE	Royal Aircraft Establishment, Farnborough
RALS	Remote Augmented Lift System
RCS	Radar Cross Section
RFC	Royal Flying Corps
RNAS	Royal Naval Air Service
RPM	Revolutions Per Minute
RRE	Royal Radar Establishment
RTO	Resident Technical Officer
RWR	Radar Warning Receiver
SABA	Small Agile Battlefield Aircraft
SAM	Surface-to-Air Missile
SHP	Shaft Horse-Power
STO	Short Take-Off
STOVL	Short Take-Off and Vertical Landing
T/C	Thickness/Chord ratio
TMR	Thrust Measuring Rig
TSR	Tactical Strike and Reconnaissance
UAV	Unmanned Aerial Vehicle
UCAV	Unmanned Combat Aerial Vehicle
UFA	Unmanned Fighter Aircraft
UMA	Unmanned Aircraft
USAF	United States Air Force
USMC	United States Marine Corps
USSR	Union of Soviet Socialist Republics
VG	Variable Geometry
V/STOL	Vertical/Short Take-Off and Landing
VT	Vectored Thrust
VTOL	Vertical Take-off and Landing

Introduction

Hawker Aircraft Ltd and its successors, Hawker Siddeley Aviation and British Aerospace plc, based at Kingston, was responsible for many of the UK's fighter and light bomber/ground-attack aircraft in the twentieth century. The fact that most of these designs emerged from a somewhat dowdy back street next to the railway in the town, rather than some impressive corporate headquarters is perhaps surprising but, in some ways, summed up the operating ethos of the company: be the best, not necessarily the first, and do it cheaply. Because the Kingston base of Hawker Aircraft housed the design team from which so many designs, including those considered here, sprang, it is right that the origin of the company and the evolution of its design department should be considered: hence the first two chapters.

All the projects described in this work were products of the Cold War. For anyone aged over 35, such a phrase has no need for further explanation; they were part of it. But for many (indeed, one fears, for most) below that age, the Cold War was something that they have vaguely heard of from their parents, something to do with Russia?

As the Second World War ended with the unconditional surrender of German and Japanese forces in the summer of 1945, the victorious Allies were looking forward to a period of quiet within which, firstly, to consolidate their victory and, secondly, to return to the ways of peace. But, for the UK, the effects of war had been devastating. As John Gaddis remarked in his seminal work on the Cold War, '[the] country was reeling from the costs of a military victory that had brought neither security, nor prosperity, nor even the assurance that freedom would survive'.

The USA had no plans to remain in Europe once the Zone of Occupation in Germany was calm. Indeed, its traditionally neutral stance meant that, politically, the withdrawal of its troops from Europe could not happen soon enough. Stalin's armies, however, were already in Western Europe and he had no intention of withdrawing them any time soon. For him, security was paramount – for him, his country and his ideology. Never again would the USSR suffer such a grievous invasion as it had suffered at the hands of Germany (that Stalin had facilitated the start of the war by agreeing with Hitler the dismemberment of Poland in 1939 was quietly papered over). To this end, Stalin sought to retain control over most of the territories that his armies occupied at the end of the war. That this included nations other than Germany was but a small concern; these countries would be offered elections which would be free and fair provided governments friendly to continued Soviet influence were elected. If not, then there were other ways of obtaining what was required.

As continuing Soviet coercion and aggression mounted in Europe, disquiet among the Western Allies grew. While the USSR was not yet strong enough to use military force to support its claims, other means were available to encourage the West to see the Soviet 'point of view'. Finally, on 24 June 1948, the land routes that connected West Berlin to the Allied zones of West Germany, were severed by the USSR, effectively isolating the Allied sector of the city in the midst of Soviet-held East Germany, the intention being to 'encourage' the abandonment of the enclave by the Western powers. As the first major provocation of the Cold War, Stalin no doubt thought that he held all the cards in this particular standoff; the USA was desperate to withdraw its remaining troops home, the UK was a spent force and France was still a shattered nation; there could be little point in the West fighting what must have seemed a fait accompli.

Yet, not only would the Allied powers keep the population of West Berlin fed and clothed with a massive airlift, the action would awaken any who still needed to see to the intentions of the Soviet Union and lead in 1949 to the formation of the North Atlantic Treaty Organisation – NATO – which would effectively bind the USA to the nations of Western Europe with economic and military aid, enabling these countries to rebuild their depleted economies and armed forces and assist in the supply of modern weaponry. Stalin eventually accepted the inevitable and re-opened the land corridor but it was too little too late; henceforth the actions of the USSR and, indeed, NATO, were viewed by the opposition as inevitably aggressive, suspicion replacing what little trust there may once have been.

With the two sides armed with the means to devastate the planet and civilised life along with it, many rightly feared for the future. Others, those who knew just how much worse things were than the public was being told, worried that there might not be a future.

Within this tense global situation, UK aircraft manufacturers were seeking some guidance from the Air Ministry as to just what it required in terms of future equipment. The answer, if any were forthcoming, was that the country would stick with the aircraft with which it had been in possession at the end of the recent war. This had the fortunate advantage of avoiding expensive new projects and any decisions on future hardware.

So it was that the British aircraft industry, upon which the requirement for new aircraft designs would fall, found itself facing a vast drop in orders, a critical lack of research funding and an Air Ministry that appeared befuddled as to what, if any, equipment should be ordered for the future. Into this vacuum, Hawker, amongst other companies, planned what they believed might be the military requirement that the government would require for the immediate future. That they broadly guessed correctly says much for the expertise that had been garnered through the lean years of the 1920s and 1930s and the desperate years of total war that had fallen upon the country in 1939. In the event, Hawker Aircraft was able to design a superlative fighter aircraft that, given the various constraints of the time, was able to provide a cogent defence against the threat from the Soviet Union as it was then understood.

All the projects described here failed – in the sense that they did not achieve service entry. All were terminated at various stages of their 'lives'; the missile-armed Hunter (despite the subtitle) did fly but got no further. P.1121 and P.1154 were at various stages of construction when they were abandoned. P.1216 reached mock-up stage but SABA and UFA were confined to the drawing board. If they had progressed to flight, there is little doubt that they would have been able performers, so it is not in the technical sphere that failure should be sought but, rather, in the political sphere where resources, requirements and political will were all lacking in one capacity or another. Did these projects fail due to successful competition within the UK? Mostly, the answer is no. Any competition for the Hunter was manifest in the Supermarine Swift which, ultimately, failed to provide the qualities demonstrated by the Hunter. P.1121 and its last iteration, P.1129, offered the RAF a fairly cheap, highly capable low-level strike aircraft able to perform most of what TSR.2 offered. That it was not taken up was in part due to irritation within the Air Ministry and RAF staff at Hawker's insistence on keeping the project active and thereby threatening their acquisition of a rather 'sexier' alternative in the shape of TSR.2.

P.1154 appeared to be the answer to a question that no one was asking: the RAF remained unconvinced of V/STOL at the time and the Royal Navy just did not see any V/STOL requirement within the service. Indeed, it threatened the continuance of the big carrier that the RN was wedded to. P.1216, Hawker

Siddeley/BAe's final attempt to get an ASTOVL aircraft into service became mired in the protracted procurement process that had by now become the norm within the UK MoD, not helped by the internal 'Cold War' being fought within BAe for the top spot in design of future projects. SABA was initially a straightforward project that could have provided the country with a potent battlefield interdiction platform able to counter the attack helicopter that was making itself felt in the localised wars that were all too often springing up around the globe. This project again became mired in prolonged discussions as to what its role should be. It could have been a useful replacement for the US A-10 Thunderbolt and this is what ultimately BAe was aiming for – to break into the US battlefield/CAS market – but this never happened, in part because the requirement was largely negated by ground-to-air missiles carried by ground forces in the battle zone.

Lastly UFA, undoubtedly an idea ahead of its time. As a platform for reconnaissance, it is likely it would have been successful; the drone market has seen an exponential growth in recent years and has formed an essential element of offensive operations for some decades. But UFA was to have been much more than that, an offensive interceptor able to take on the best that the Warsaw Pact could pitch against it in aerial combat almost completely autonomously. In the last years of RAE Farnborough, work was in hand to determine the requirements for fighter aircraft wherein the pilot was an option rather than a necessity but, as far as is known, this ability has still not been achieved at the time of writing.

BAE Systems, the UK successor to the HSA and BAC legacy continues today to be a potent player in the global defence market, its air systems headquarters now centred on Warton in Lancashire. Over the past two decades the company has added land and sea systems to its portfolio and emerged in the USA as a major defence player – not bad for a UK company that was once seen as a rundown also-ran in the global market.

Over the period covered by this work the name of the company went through multiple changes; originating as Sopwith Aviation in 1912, it became HG Hawker Engineering in 1920 before changing to Hawker Aircraft in 1933. This remained the case until the creation of Hawker Siddeley Aviation in 1963, a name retained until nationalisation in 1977 when it became part of British Aerospace. In 1999 this finally became BAE Systems, the title it retains today (2023). However, to avoid confusion, the company at Kingston is generically referred to as Hawker in the text.

Chapter 1

Hawker Aircraft, Origins and Development

Hawker Aircraft Ltd of Kingston upon Thames was arguably the most successful UK fighter design house of the twentieth century. From its inception in 1920 to its amalgamation into British Aerospace in 1977, the company's products were seldom absent from the armed services' order of battle. But for every successful design, there were many others that did not reach fruition: this work examines some of these. To better understand the projects described later in this work, it would be profitable to examine the origins of the organisation from which these designs flowed and the men who were responsible for them. Ostensibly, it had its beginnings in 1920 but, as we shall see, its roots were rather deeper.

On 15 November 1920, papers had been submitted registering the formation of the HG Hawker Engineering Co. Ltd with capital of £20,000 in £1 shares, the new company being incorporated on 31 December 1921. Its Articles of Association described the proposed activities of the new company as 'the manufacture of motor cycles, and to carry on the business of manufacturers of and dealers in cycles of all kinds, infernal [sic] combustion engines and steam engines, motor cars, aircraft etc.' The first directors were announced as Frederick Ibbotson Bennett, engineer; Harry George Hawker, aeroplane pilot; Thomas Octave Murdoch Sopwith, engineer; Frederick Sigrist, engineer; V.W. Eyre, engineer.[1] By this time, aeroplanes had come a long way from the trial-and-error experiments of the first decade of the twentieth century; their design was now subject to established guidelines based on experience and science, rather than individual foible. On what basis then did this fledgling company claim to be in a position to manufacture the latest aircraft? What experience and knowledge resided in those founders? Did they indeed, know *anything* about the business of aircraft design? Their decision to begin an aircraft manufacturing concern in a market awash with unwanted surplus military aircraft from the recent world war certainly did not augur well for the business sense of those involved. So, just who were these 'businessmen'?

T.O.M. Sopwith was born on 18 January 1888, the son of a wealthy civil engineer. As a young man he had been attracted to the forefront of engineering expertise in the form of motor cars and cycles, fast boats and balloons and was a contemporary of C.S. Rolls who would later found the Rolls-Royce company. In 1910, Sopwith turned his interest to powered flight following an initial trip aloft at Brooklands, the first dedicated banked motor-racing circuit in the UK at Weybridge in Surrey, the inner grass expanse of which the owner Hugh Locke King, had allowed the use of by the early aircraft pioneers. Thereafter Sopwith purchased his own aircraft – a Howard Wright monoplane, with which he taught himself to fly and soon received certificate number 31 from the Royal Aero Club. Having entered two early competitions for flight duration and won that for the longest flight from the UK to the continent, Sopwith toured the United States giving displays and meeting similarly minded contemporaries; his initial plan had been to complete a crossing of the continent from coast to coast but dropped the idea once in country. Back in England, now with several aeroplanes, he decided to set up his own school of flying at Brooklands, the Sopwith School of Flying opening in February 1912.[2]

In an interview with John Crampton late in his life, Sopwith recalled those early days. 'I was running a flying school at Brooklands at the time, teaching people to do something I knew very little about myself. The Royal Flying Corps was formed in April 1912 and later that year I was approached by a Major in the Scots Fusiliers, with a deep booming voice, who wanted to join the RFC. Unless he held an aviator's certificate within the next ten days, he would be over-age to join and he asked me if I could get him through his tests in time. He was successful, "Boom" Trenchard, who was to become the Father of the Royal Air Force.'[3]

Harry Hawker was an Australian, born in 1889. He had a great interest in engineering and travelled to the UK in 1911 to further this interest. At first employed in a variety of motor-car related jobs, he obtained a position with T.O.M. Sopwith the following year and learned to fly with Sopwith's school at Brooklands. Sopwith could now employ him as a pilot as well as a mechanic and Hawker quickly took on the role of testing and demonstrating the various products from the nascent Sopwith company. His qualities as an engineer and pilot allowed him to become involved in the design aspects of Sopwith aircraft, leading to an influential position at Sopwith Aviation which he retained into the post-war years.

Frederick Sigrist was another of Sopwith's early associates, employed by him in 1909 when Sopwith acquired a share in a yacht and it was Sigrist, an intuitive rather than educated mechanic, who was responsible for its upkeep.

With Sopwith's increasing interest in aviation, Sigrist followed his employer into this field, maintaining the flying school machines and then suggesting improvements to them. With the formation of Sopwith Aviation, Sigrist took on an influential role in design and modification and, with Sopwith and Hawker, became part of the team responsible for subsequent Sopwith designs. At Sopwith Aviation, Sigrist had been Works Manager and with HG Hawker Engineering Ltd assumed the position of Managing Director.

William Eyre was yet another early acquaintance of Sopwith, it being he who owned the other half of the boat that Sopwith had acquired in 1909. Little is known of F.I. Bennett but he was another Works Manager and may have operated as the Sopwith Company Secretary since, at the formation of HG Hawker Engineering, it was Bennett in whom the various Sopwith patents and designs were vested.[4]

Out of Sopwith's flying school activity came the desire to modify existing aircraft and then to design entirely new ones. With a small staff, this work

T.O.M. Sopwith aloft in his Howard Wright monoplane at Brooklands, c.1911-12.

soon produced what he called a Sopwith Wright biplane capable of carrying two passengers and he was able to sell the machine to the Royal Navy, an outlet which would later pay dividends. Following the relative success of this design, his interest in the flying school diminished as that of design and construction of new aircraft increased. Premises were obtained in Canbury Park Road, Kingston upon Thames, Surrey, as a base for the creation of the Sopwith Aviation

The roller skating rink in Canbury Park Road, Sopwith Aviation's first premises in Kingston upon Thames, 1909.

Company Ltd, with Brooklands available for erection and flight testing of new designs. These first premises, the former roller-skating rink in the town, given its large open covered space and level wooden floor, proved admirably suited to the early work and by 1913 an improved three-seat biplane – the 80hp D1 – (thirteen built, first flown 7 February 1913) and a flying-boat were produced which attracted much interest at the Olympia Show of that year. The early relationship with the Naval Wing of the RFC proved fruitful, the company receiving several small orders for flying boats.

Around 1913 the company produced what would become the basis for a number of biplane designs, called the Tabloid (twenty-seven built, first flown 27 November 1913) and a float-plane version of this was used to win the Schneider Trophy in 1914. However, it was the advent of the First World War that would bring the company to prominence. Large orders for aircraft started to arrive for both the naval and the military wing of the RFC, including the 1½ strutter, of which 1,435 were built for the RFC and RNAS plus 4,500 manufactured by French companies for the French Air Force. This was followed by the Pup,

Sopwith Tabloid at Brooklands with T.O.M. Sopwith and Harry Hawker, c.1913.

1,846 built; the Triplane, 152 built and the first operational aircraft with this wing configuration, much praised and copied by the German forces; and that most famous and widely produced fighter of the war, the Camel with 5,996 produced; ending the war with the Dolphin of which 1,775 were constructed and the Snipe, 2,097 built. Most of these designs were those of Herbert Smith, Sopwith's designer from 1914.[5]

Smith had taken a diploma in engineering at Bradford Technical College and had later gained employment with the Bristol Aeroplane Company before moving to Sopwith in 1914 as a leading draughtsman. From late November, Smith took on the role of Chief Designer, being responsible for the Triplane, Snipe, Dolphin and many others which did not see production contracts.

In 1917, the government, in an attempt to rationalise the production of aircraft, had produced a plan to create state-owned factories for this purpose, the first of which (Aircraft Factory No.2) was erected north of Kingston at Ham. Asked to take on the operation of this for the government, Sopwith declined but agreed to its lease and used the premises to expand production to far greater capacity than was possible in the Canbury Park Road premises. By the end of the war, over 16,237 Sopwith aircraft had been produced by the parent company and its sub-contractors, (4,133 at Brooklands) a remarkable example of efficient production technique and management (11,237 in the UK, 5,000 under licence in France and Italy. This was from an advertisement for the Sopwith Aviation and Engineering Company Ltd in *The Aeroplane*) T.O.M. Sopwith would go on to figure large in the fortunes of the Hawker Aircraft Company as will be described later. With the dissolution of Sopwith, Herbert Smith did not stay with the new company for long, instead, in 1921, taking a role in Japan with Mitsubishi to modernise their aircraft industry, a role that was rather too successful as the successors of his designs were at the forefront of the Japanese aggression at Pearl Harbor and in South-East Asia.

With the signing of the Armistice on 11 November 1918, military contracts began to be cancelled and, almost at a stroke, the majority of Sopwith Aviation's work ceased. By early 1919 the company had been renamed Sopwith Aviation and Engineering Ltd and sought to undertake general engineering manufacture while seeking new designs with which to pursue what little market remained for aircraft in the post-war years. Having divested itself of the Ham works and retrenched into the Canbury Park Road enclave, the company was set to eke out a meagre existence while waiting for the aviation market to recover. All might have been well even though the company was losing money but in 1920 a claim

Sopwith's Aircraft Factory No.2 and later Hawker's HQ at Ham Common, Kingston upon Thames, in December 1918 with Snipes, Dolphins and Salamanders under construction.

from the Treasury for Excess War Profits Duty, lodged against the company, proved the final straw. This large bill had been payable each year after 1915 and was duly paid by the company but, with the cessation of the war and the loss of contracts, the duty continued to be claimed until 1921. Although Sopwith had been able to pay its tax liabilities during the war, it could not do so postwar while continuing in business. The decision was made therefore to place the company into voluntary liquidation, pay off creditors and the Inland Revenue and start afresh.[6]

Thus, on 15 November 1920, the HG Hawker Engineering Company L was born, taking the name of Harry Hawker, rather than Sopwith, to prevent any claim that it was the same company. That said, HG Hawker Engineering (HGHE) occupied part at least of the old Sopwith premises in Kingston, retained largely the same workforce and sought to pursue the same general engineering concerns as the old company, including aircraft.

Sir Thomas Octave Murdoch Sopwith CBE, founder of the Sopwith/Hawker/Hawker Siddeley aircraft companies.

While the old company (Sopwith) had utilised the design skills of T.O.M. Sopwith, Fred Sigrist and Harry Hawker, and later, Herbert Smith, it was decided that, once aircraft work picked up again, with Smith's departure to Japan, a dedicated designer/draughtsman would join the team, that person being Captain B. 'Tommy' Thomson RN, employed in 1922. His employment may in part have been prompted by the untimely death of Harry Hawker while practising for the Aerial Derby in July 1921, probably as the result of a haemorrhage due to tuberculosis following a crash caused by engine fire. However, the company weathered the storm and, retaining Hawker's name, began to acquire small contracts for complete refurbishment of aircraft, including their own Snipe. Captain Thomson started to produce new designs, the first being a reconnaissance parasol monoplane called the Duiker and a biplane night-fighter called the Woodcock. Both proved to be poor designs; the Duiker was soon consigned to the bin followed shortly by Thomson. His place was taken by Wilfred George Carter who was able to rework the Woodcock and produce a more acceptable aircraft. Carter had been Chief Draughtsman with Sopwith Aviation so had a good knowledge of the new company and its management which no doubt had confidence in his abilities.[7]

In November 1923 a new employee joined the ranks at Canbury Park Road. While his arrival may not have been particularly noted, this man would go on to lead the design of Hawker products for the next forty-three years and, in the process, establish Hawker as the leading UK fighter design house of the twentieth century. Sydney Camm, born in Windsor, Berkshire, in August 1893, had left school in 1908 to take up a carpentry apprenticeship. With his brothers, like many boys then and since, he developed an interest in aircraft and aeronautics in general, making model aeroplanes which they sold in the locality. In 1912 he was one of the founding members of the Windsor Model Aeroplane Club, an outlet that allowed him to develop his understanding of aircraft construction and aerodynamics. Around the beginning of the First World War, Camm had secured a position as a carpenter with the Martinsyde Aircraft Company (formed by H.P. Martin and George H. Handasyde) at Brooklands, Weybridge, before moving into the Drawing Office, but at war's end, the company, like many other aircraft concerns, went into liquidation and, in the early 1920s, Camm moved to premises acquired by George Handasyde at Mayford, Woking, before moving on to the fledgling Hawker Engineering Company in Kingston upon Thames in 1923 as a senior draughtsman. With Carter's move from Hawker to Gloster Aircraft in 1924, from 1925 Camm became Chief Designer at Hawker and began design of a long line of successful aircraft.[8]

As far as aircraft design was concerned the early 1920s were really a continuation of the construction techniques developed prior to the First World War and refined during it. The primary requirements for these aircraft (in particular the military variety) were speed, manoeuvrability and endurance and, one might add, a stable platform for weapons use. Speed was a function of engine power and low drag design. The engines used during the war had been principally the rotary type where the engine rotated around the fixed crankshaft, but the post-war period saw these replaced with the more fuel-efficient radial type. Later in the 1920s, however, the use of in-line engines with their liquid-cooling systems offered the prospect of superior streamlining, allowing greater speeds to be achieved. It was the greater efficiency and power of the engine that would govern the success or otherwise of aircraft for many decades to come. However, manoeuvrability came from the aircraft design itself. Weight was the single most important consideration when designing aircraft in this period; of course, it was important to be able to actually leave the ground but light weight would offer improved speed and manoeuvrability in the air, allowing the pilot to dominate any action in which he was involved. To achieve this light weight, the structure was constructed as a wooden skeleton which was then covered in fabric to achieve the aerodynamic surfaces required for flight. This structure was braced internally with wires to give a strong, light fuselage capable of carrying engine, pilot and whatever offensive load might be required. The wings were similarly constructed, the almost universal use of the biplane configuration after initial flirtation with monoplanes, giving a well braced, rigid structure which would resist deflection when manoeuvring and flutter when diving; the low wing loading also allowed superior manoeuvrability.

It will be seen from this that the primary trade upon which the construction of aircraft depended was that of the carpenter who would have the knowledge to select the strongest and lightest materials and the ability to shape them to best effect. Of equal importance was the engine mechanic, able to master the internal combustion designs and get the best from them. Engines of this period were renowned for packing up in flight, something that the pilots must always have had in the backs of their minds though it seldom seemed to diminish their enthusiasm. Perhaps the other important trade was that of rigger. The entire construction was of no use if it was not rigged correctly; the bracing wires had to be carefully tensioned to hold the aircraft in a rigid configuration with the wings at the correct angle of dihedral and incidence. Any movement of the wing structure could quickly lead to vibration and flutter and the loss of a wing, usually with unfortunate results.

The use of wood had its limitations in terms of strength and weight. As design moved into the 1930s, aircraft were increasingly constructed using a

Headquarters of Sopwith Aviation Ltd and later, HG Hawker Engineering Ltd and Hawker Aircraft Ltd in Canbury Park Road, Kingston upon Thames.

metal framework which gave superior strength for a given weight. This led to woodworking skills becoming less and less of a requirement and the ascendancy of the skills of the metal-worker, a situation which would endure through the rest of the century.

During the period from the formation of HG Hawker Engineering to the start of the Second World War, the design of aircraft changed greatly, yet, to the casual eye, there might not have been much difference between one designed in 1920 and that designed in 1934, such was the conservatism prevalent in the aircraft industry driven by that within the Royal Air Force and Air Ministry. The standard design was that of a biplane with one- or two-bay wings, fixed undercarriage, fabric-covered flying surfaces and simple armament aligned to fire forward in the pilot's line of sight. The pilot sat in an open cockpit with his

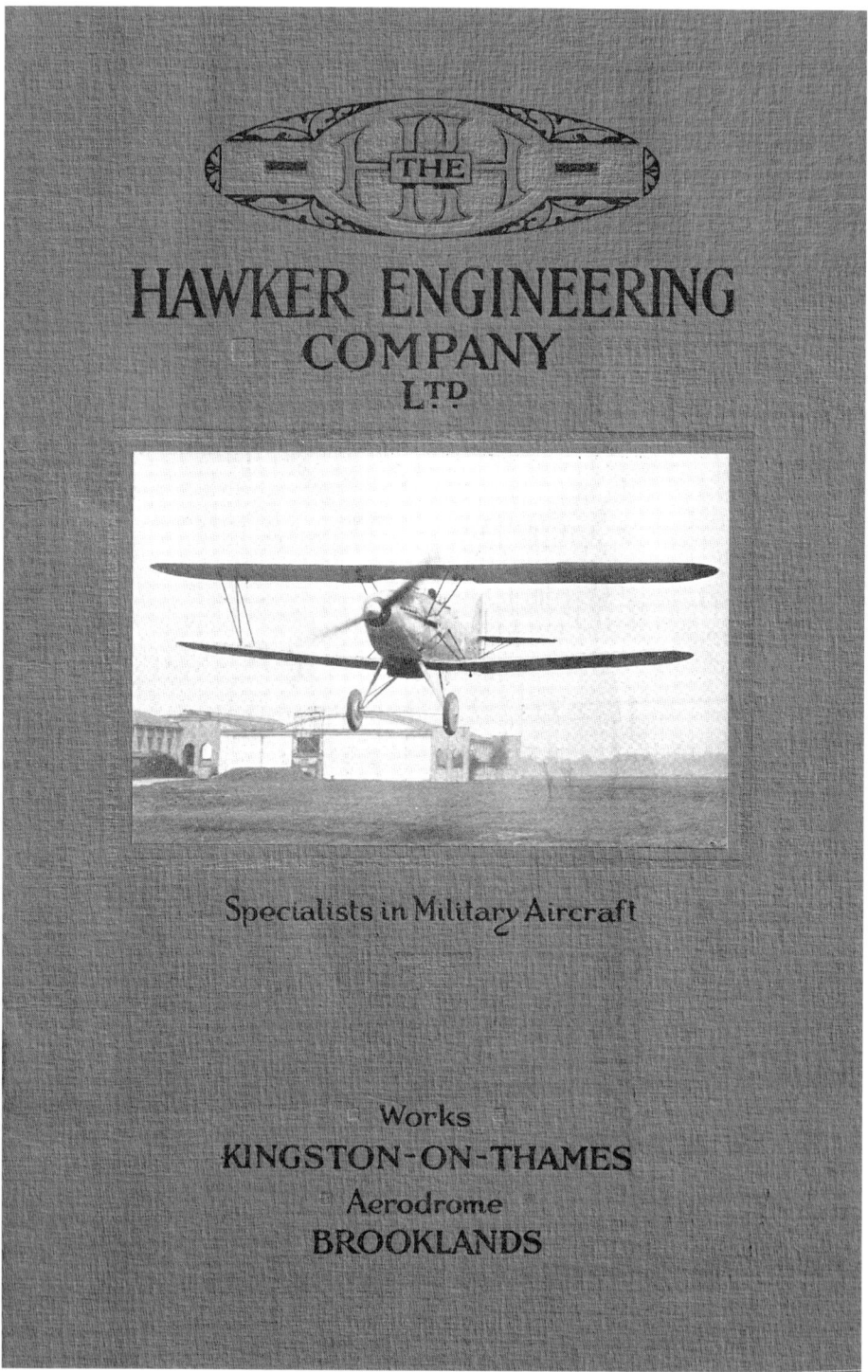

Brochure advertising the work of HG Hawker Engineering Ltd, c.1920s.

view restricted by the wings, the large engine and the guns in front of his face. Not until the prejudice within the higher echelons of the RAF and its ministry against the monoplane was overturned could the design of aircraft proceed at a greater pace.

At Hawker, design, as led by Camm, was seen as an iterative process. There was little to be gained from radical leaps in technology, even if that had been possible. The process was more akin to small improvements introduced over a series of different designs (and there were plenty of these: thirty-seven up to the Hurricane). One of those improvements was the replacement of the internal wooden frame by one of metal construction. At first, the various members were welded together to form the framework, Fred Sigrist being a keen welder. However, Camm devised an alternative scheme whereby the framework was comprised of a series of round or square section tubes, their ends swaged to a rectangular section, which instead of being welded at the joints, were bolted together with plates. The advantages of this method included the ease with which damaged sections could be replaced, in the field if necessary, and the fact that relatively unskilled labour was required for the job instead of welding ability. This method was progressively introduced across the Hawker designs, beginning with the Hornbill of 1925, so that by the time of the Heron the following year the entire internal structure was created in metal. Of course, this did not preclude the continued use of bracing wires. Their deletion would have to wait.

Another innovation was the so-called dumbbell spar which comprised two polygonal steel beams held one above the other by a metal web. The result produced a very strong but low weight main spar for the wing. By the time of the Hart series of aircraft, Hawker aircraft structures were all-metal with sheet metal covering the fuselage from nose to cockpit and fabric-covered wooden formers overlaying the metal construction aft of this. The metal flying surfaces – wings, tailplane, elevators, fin and rudder – were still fabric-covered. Thus did design proceed through the 1930s, though it could sometimes look as if the 'design' chairs were merely being re-arranged. Greater engine power and improved streamlining, particularly benefitting from in-line engines, allowed for a gradual increase in speed, each new model offering some small improvement. However, it was clear to Camm, and indeed to other designers, that any further increase in speed would have to come from abandonment of the biplane design.[9]

To this end, Camm had, in 1933, begun to toy with a new monoplane design, modelled on the Fury biplane fighter. This design, which would eventually

Premises of Hawker Aircraft Ltd at Brooklands showing the assembly sheds against the motor racing track banking and the flight hangars in the foreground.

find fame and fortune as the Hawker Hurricane, would feature a number of innovations as well as the monoplane layout. A fully-enclosed cockpit would offer the pilot the shelter from slipstream that the higher speeds being attained made important. A retractable undercarriage would allow this draggy piece of the aircraft to be tucked away, further enhancing performance. The new engine from Rolls-Royce – the PV.12 – would mature into the Merlin series of engines, eventually giving improved performance in a concentrated unit that allowed for close cowling and less drag. Perhaps most importantly for its designation as a fighter was the location of massed guns in the wing firing forward outside of the propeller disc, initially eight but later increased to twelve and later still, a four-cannon armament.

With a new specification F.36/34 written around Camm's latest ideas, a contract for one example of a 'High Speed Monoplane' was issued in February 1935. As the Hurricane entered production, its construction looked forward to the aircraft of the future with its cantilever wing, powerful inline engine, enclosed cockpit and retractable undercarriage, but also backward, to the tried-and-trusted Hawker metal-framed structure and fabric-covered flying surfaces of the biplanes. Even before the outbreak of the Second World War, Camm was scheming the Hurricane's replacement which would culminate in the Typhoon fighter. For the first time, Hawker would alter its design technique to incorporate monocoque construction in the new aircraft aft of the cockpit while retaining the traditional metal framing forward of the cockpit.[10]

As noted above, the 'Hawker' method of construction had used a skeleton comprised of tubular members which took the stresses and loads on the airframe while the skinning had no load-carrying requirement. In monocoque construction, however, the internal tubular construction was dispensed with, the load paths instead being borne by the stressed-skin construction built up over a structure of frames and longerons. This method *potentially* offered greater rigidity and strength to combat the increasing loads imposed on modern aircraft but was far more complex to manufacture. The Supermarine Spitfire had used this construction method from the outset, which was one reason for the long delay in getting useful numbers of production machines to the services in time for war.[11]

Whilst it may be assumed that monocoque construction was a recent innovation, in fact it had been used at an earlier date and had featured on aircraft of the First World War, including several Sopwith designs such as the Snail and the Snark. However, it created a heavier structure than the fabric-covered variety. Quite why Camm decided to split his options on which construction type was used in the Typhoon (and the Tornado) is unknown; perhaps his innate conservatism was at work again. As it happened, severe problems with the Typhoon soon arose when the prototype aircraft began to suffer control difficulties. Philip Lucas was, with difficulty, able to get the aircraft back to Langley and land, whereupon it became evident that the monocoque rear fuselage had failed behind the cockpit. Having addressed this weakness, further flying revealed a predilection for the tail unit to part company with the aircraft at the rear transport joint when flown at high speed, inevitably resulting in tragedy and calling into question the integrity of the design. After much investigation, the problem was traced to a small bracket on the elevator mass balance, which was failing, allowing severe vibration to occur leading to a load failure at the transport joint, with monocoque construction getting a clean bill of health.

Camm's next design to enter service, the Tempest, was perhaps the best fighter in the latter stages of the war, and again used stressed-skin construction in the rear fuselage. A development of the Tempest, the Fury and the Sea Fury, would mark the introduction of all-monocoque construction and the end of Hawker designs powered by reciprocating engines and propellers; henceforth, company aircraft would feature the new gas turbine as their prime mover. The first jet design to reach manufacture was the P.1040, which would go on to be produced as the Hawker Sea Hawk for the Royal Navy. In this design, many features novel to the company were introduced: jet engine power, wing-root air intakes, tricycle undercarriage, split-jet exhaust and the highest speeds yet attained by a Hawker aircraft. However, the installation of a jet engine could bring its own design challenges. Formerly, the CG was governed by the great weight of the engine being mounted in the nose of the aircraft. This resulted in the pilot being seated behind the engine leading to a restricted forward view, especially on the ground when taxiing. The introduction of jet engines meant that they needed to be positioned much further back, being located behind the pilot on the longitudinal centreline and the pilot position then moved forward to the very front of the aircraft. Once this reversal of normal CG challenges was overcome, this became the arrangement of choice and gave the pilot an excellent field of view unencumbered by large engines directly in front of him. But by now we are into the post-war period. Having examined the various innovations introduced into Hawker designs in this period, it would be useful to examine the structure of the design organisation within the company.

It should be mentioned here that, prior to 1939, Hawker's entire design and manufacturing process was undertaken in the centre of the busy market town that was Kingston upon Thames. The not particularly salubrious premises were housed in the equally insalubrious surroundings of Canbury Park Road. In the 1920s the Drawing Office was located on the first floor of the office frontage. As the 1930s dawned, this building was extended and floors added, eventually being three floors throughout. Running back from this office complex ran a series of further buildings conjoined to form a large 'island' site incorporating the various manufacturing and assembly shops, including a dope- or paint-shop which also dealt with the fabric covering of wings and fuselage. On the south side of the road, a new three-storey building was constructed in 1933/34 to accommodate the Experimental Shop where the prototype Hurricane was constructed. Adjacent to this building was Bentall's furniture depository which, in 1937, was taken over by Hawker with the Drawing Office moving into it.

Factory and airfield of Hawker Aircraft Ltd at Parlaunt Park, Langley, Bucks, c.1940/45.

The new design offices did not spoil the staff, used as they were to the industrial outlook from the old office; they were now afforded an unparalleled view of the railway line and Bentall's new depository. Lastly, also on the north side, Hawker retained the original Sopwith building which had started life as a roller-skating rink. Thus, by the time of the Second World War, most of Canbury Park Road was given over to Hawker Aircraft Ltd. This ad hoc growth of the company in the centre of a major town resulted in a cramped, inefficient environment, the antithesis of what was required for volume production of aircraft and, because of this, much production effort was sub-contracted out to other companies within the Hawker Siddeley Group.[12]

    The absence of an airfield upon which to test the new aircraft was another factor in the competitive market in which Hawker operated. Unlike other companies, Hawker could not wheel their new product outside and put it through its paces. As each aircraft neared completion at Kingston, the fuselage and wings would be transported 10 miles down to Brooklands airfield where final assembly and flight took place, the Hawker flight shed being located in two Belfast-trussed hangars within the motor-racing circuit left over from the RFC presence during the First World War. To these were added, around 1937, further assembly sheds up against the racing circuit banking within which to complete assembly prior to first flight. This dispersed 'ad-hoc-ery' did nothing for production efficiency but was entirely of a piece with the hand-to-mouth early days of Hawker Aircraft and it is likely that, without the intervention of the threat from Nazi Germany, nothing would have changed until change was forced upon the company.

In 1933 T.O.M. Sopwith had led plans to expand the company's manufacturing base by acquiring other of the UK's aircraft manufacturers, many of which were struggling in the 'disarmament' atmosphere then prevalent in the country and reflected in the governments of the day. To this end, that year, the company name of HG Hawker Engineering (valued at £909,367 7s 8d) was replaced on 1 April 1933 with a private limited company – Hawker Aircraft Ltd – and in 1934 the complete assets of the Gloster Aircraft Company were acquired and initial contracts placed with Gloster for the manufacture of Audax and Hardy fighters. Later, Gloster would build Henley, Hurricane and Typhoon aircraft

Airfield of Hawker Aircraft, Hawker Siddeley Aviation and British Aerospace at Dunsfold, Surrey, c.1995.

for Hawker, placing Gloster on a firm footing for many years. In 1935 Sopwith announced the formation of a trust to acquire all the shares of the Armstrong Siddeley Development Company, to be held within a holding company named Hawker Siddeley Aircraft Company. Thus was formed a powerful grouping of UK aircraft companies with which to face the growing threat from Nazi Germany and which comprised Hawker Aircraft Ltd, Gloster Aircraft Co. Ltd, Armstrong Whitworth Aircraft Ltd, Armstrong Siddeley Motors Ltd, A.V. Roe and Co. Ltd (Avro) and Air Service Training Ltd.[13]

As the later 1930s witnessed the continued growth of the threat from Germany, the government of the day was eventually prevailed upon to turn its back on the disarmament policies of earlier years and begin a series of re-armament moves which would fall principally on an aircraft industry brought to the brink of collapse by earlier policies. The UK would now look principally to its aircraft companies to provide the weapons with which to oppose the ever increasing threat from the continent. Part of this growth was to be created by the erection of 'shadow' factories, vast sites which would be able to churn out the aircraft of the day, often sited adjacent to volume car producers, who were expected to manage the sites and provide much of the labour, but the core would be the aviation companies that had struggled through the lean years of the 1920s and early 1930s.

As Correlli Barnett observed so aptly, 'On the eve of re-armament in 1936, therefore, the British aircraft industry remained a cottage industry with obsolescent products, sleepy firms with little more than experimental aircraft shops employing hand-work methods and centred on their design departments'. What was needed was, in essence, a new industry capable of volume production using modern machine tools and mass-production methods.[14]

In the case of Hawker Aircraft Ltd, the requirement for large-scale expansion was to be answered by a new purpose-built facility, although under Hawker control. To this end, land was purchased at Parlaunt Park Farm, Langley, Bucks, in 1936 for the construction of a large factory and aerodrome. Opening for business in 1939, Langley would go on to produce Hurricanes and Tempests (most Typhoons were built by Gloster) as well as the Sea Fury after the war, the aerodrome finally closing in 1959 due to the lack of hard runways for the new jets being constructed and the clash with air traffic from the new Heathrow Airport. Henceforth, all flight testing would take place at Dunsfold Aerodrome, acquired in 1951, thus mirroring the 'dispersed' nature of the company in the days of Brooklands. Also acquired, or perhaps one should say 're-acquired', was the Richmond Road factory at Ham Common vacated by Sopwith in 1920. Leyland Motors moved their operations north in 1948 and the site was purchased by Hawker Aircraft in time to begin manufacture of their new jet fighters now coming to the fore. Eventually, the company would vacate the Canbury Park Road premises, although not until the early 1960s, and concentrate its resources in Richmond Road where, in 1958, a large and imposing suite of offices was erected fronting the Richmond Road and obscuring the manufacturing areas from sight. With this construction work complete and Dunsfold Aerodrome developed and secured, little in terms of infrastructure would change till the end.[15]

Also resistant to change would be the fundamental nature of the workforce available to Hawker and the industry in general. Much of this was the result of what Correlli Barnett termed the 'practical man', the result of 'on-the-job' or in-house training as shop boy or apprentice. What was missing was any technical education acquired during the individual's schooling such that there was any basis on which to build advanced technical knowledge (Sir Sydney Camm might be seen as the epitome of the 'practical man' at high level). This was especially noticeable with the requirement for technical design and development work in the industry, notwithstanding the influx into the Kingston Project Office of graduate engineers such as Hooper and Fozard after the war. The relatively straightforward skill requirements of the tubular framework of earlier Hawker aircraft had masked this deficiency but, as monocoque construction became the order of the day, it became more noticeable.[16]

As the post-war aircraft industries of the USA, Germany and France raced away, that of the UK struggled to compete in an ever more crowded market. From the illusion of the UK as a first-class world power in the 1930s had come the realisation that, in 1946, it was merely a dowager/pensioner, kept afloat by the financial might of the USA. If the post-war period confirmed the UK's position as pauper state, the US, facing the same economic problems in the 1930s as the UK, had been able to emerge into the mid-1940s in an almost unassailable position, thanks significantly to massive orders from the UK to the US military-equipment industry. The UK's failure in post-war exports lay not so much in the lack of natural and manpower resources that it possessed as in the technical abilities of that manpower due to the paucity of technically-educated labour available to business and industry, a situation addressed much earlier in the economic histories of other nations, especially Germany.

Be that as it may, what would change, and radically, would be the nature of the management and control of the company. With the further acquisition

Headquarters of Hawker Aircraft, Hawker Siddeley Aviation and British Aerospace at Ham Common, Kingston upon Thames.

of aircraft companies by the Hawker Siddeley Group board in the early 1960s (Folland Aircraft, Blackburn Aircraft and de Havilland Aircraft), the group was re-organised into two operating units – Hawker Siddeley Aviation and Hawker Siddeley Dynamics. Along with the other aircraft manufacturing companies, Hawker Aircraft was grouped into 'Aviation' while the various guided-weapons concerns of the companies were grouped under 'Dynamics'. In this way, Hawker Aircraft ceased to be master in its own house. Henceforth, from 1 July 1963, Hawker Siddeley Aviation, as it was now known, would have to share control and funding with its former competitors as will become clear during discussion of the P.1121 saga.

As the newly-constituted group settled down with its new partners and learned to get along amicably, Hawker Siddeley (Hawker Blackburn Division) got on with the job of producing winning designs for the armed forces of the UK and anyone else deemed a suitable customer by HM government for their products abroad. The 1960s and 1970s saw the industry becoming more and more enmeshed in the politics of the day, culminating in the disaster that was the TSR.2. Sir Sydney Camm summed it all up in one of his infamous epigrams: 'Aircraft of the future will have to possess a fourth dimension – politics!'

Hawker Siddeley Aviation would remain a decisive force in UK aircraft design until, in 1977, the nationalisation of the aerospace and shipbuilding industries by the Labour government produced a new entity – British Aerospace, subsuming Hawker Siddeley and BAC into a new state-run monolith, resulting in the division being renamed the Kingston-Brough Division. Although design and manufacture of aircraft at Kingston would continue within the new monopoly until closure of the site in 1992, management control would slowly ebb away from Kingston and Weybridge to the new northern stronghold of Warton, the former English Electric site at Blackpool.

Thus, in 1992, production of aircraft in Kingston ended after over eighty years and of Hawker products seventy-two years, leaving Kingston's flight-test centre at Dunsfold the one remaining beacon of the old Hawker enterprise until 2000 when this, too, was closed. Not a bad run from a company whose first aircraft were designed in an old roller-skating rink.

Chapter 2

The Hawker Design Office

In the earliest days of powered flight, it is important to realise that there were no hard-and-fast rules governing aircraft design. The primary requirement, of course, was to construct a machine capable of leaving the ground with a pilot, no easy task in itself. Whilst the design of the Wright Brothers' first successful aircraft and their subsequent tour of Europe influenced British design, the individuals involved in flight in the UK were certainly not slaves to the Americans' ideas. Samuel Cody, for example, (actually an American though experimenting in England) had begun his trials with man-carrying kites under the auspices of the Royal Balloon School at Farnborough before these ideas metamorphosed into something more akin to an aeroplane. In France, Blériot eschewed the complex biplane, tail-first layout and produced a successful monoplane with the classic layout we now associate with most basic aircraft.

During the tour of Europe undertaken by the Wright Brothers, a number of licences were allotted to individuals to construct aircraft based on the Wright formula and it was one of these – a 40hp Howard T. Wright (no relation to the Wright brothers) monoplane that T.O.M. Sopwith first acquired with which to begin the process of learning to fly. This was followed by a Howard T. Wright biplane powered by a 60hp engine. Later, a Burgess-Wright biplane was acquired in the US, joining a Blériot and a Martin & Handasyde Type 4B. With these various aircraft, used to provide the beginnings of a flying school at Brooklands, Sopwith and Fred Sigrist soon learned the intricacies of the various designs, particularly as they often suffered damage and required repair and sometimes reconstruction. In Sigrist's capable engineering hands, and with input from Sopwith and his pilot Harry Hawker, the Sopwith company was able successfully to launch many designs taken up by the armed forces during the First World War and, come the post war retrenchment and the creation of HG Hawker Engineering, that expertise was immediately available to begin work on new designs for the post-war period.

The activities of the Royal Aircraft Factory, as it was then known, at Farnborough in the early war period and their design and construction of aircraft for the fledgling Royal Flying Corps (RFC), formed in 1912, caused friction with the private companies attempting to supply aircraft to the fighting services; the accusation that Farnborough's organisation was becoming something of a government monopoly was not far from the truth. T.O.M. Sopwith was among those private constructors seeking to restrict the work of Farnborough to research and leave the quantity production of aircraft to private industry, a battle won in 1916 when Farnborough was instructed to cease design and development of aircraft and restrict its activities to research and advice and to make this available to the UK aircraft companies.[1]

Following the creation of the Hawker company, it was decided to employ a dedicated designer/draughtsman and Captain Bertram 'Tommy' Thomson was retained as Chief Designer in 1922. His first two designs comprised the radial-engine, parasol-winged Duiker, a two-seat reconnaissance aircraft and the Woodcock, also radial-engined and designed as a night-fighter biplane. Neither design was considered particularly good, the service test establishment at Martlesham Heath being less than enthused by their basic design flaws. Perhaps because of this, Thomson departed and was replaced by Wilfred 'George' Carter, a rather more promising designer who took the Woodcock design and redrew it to produce an aircraft that was good enough to obtain sales to the RAF as the Woodcock II. Slowly, a small design department began to grow at Canbury Park Road and, before long, one Sydney Camm arrived to join the company as a senior draughtsman in 1923. With Carter's decision to leave Hawker and join Gloster Aircraft, Camm came to prominence and in 1925 was appointed Chief Designer.[2]

Around this time, Sigrist and Camm perfected the process of forming metal aircraft structures using a framework of tubes bolted together with plates at the joints, this method being used in most Hawker designs for the next fifteen years. Various new designs won increasing favour with the RAF and Royal Navy, the Kingston works expanding during the 1920s and 1930s as more and more property was acquired in which to construct the products of the company and, as seen in the previous chapter, eventually a completely new site would be acquired at Langley upon which to build a new factory and aerodrome.

As time passed, the design team that would stay with Camm for the next decade and more slowly formed. In 1926 Roland Henry 'Roy' Chaplin, joined the team, quickly showing his design acumen and becoming influential in the design process, being particularly associated with the dumbbell spar discussed earlier. However, it would seem that the design in fact originated with the Sir W.G. Armstrong Whitworth Company since, in November 1927, a licence

The Hawker Design Office   25

Sydney Camm as a young man, with one of the models created at the Windsor Model Aeroplane Club.

agreement was concluded between that company and HG Hawker Engineering Co. Ltd for spars to the former company's designs including 'the double tubular boom construction in which two booms of circular or ovalised section are separated and connected by single or double webs'.[3]

He would become Camm's right-hand man as Deputy Chief Designer and oversee many of the projects to come, eventually becoming Chief Designer on Camm's death. In these early days, the Design Department – then known as the Drawing Office – comprised fewer than forty people, including tracers, print-room staff etc. Also joining the company in June 1927 was Harold Tuffen as a junior draughtsman; a decade later he was promoted to be a section leader

Sir Sydney Camm CBE, Chief Designer, Hawker Aircraft Ltd.

in the Experimental Drawing Office dealing with project design, including a period at Claremont on new types' project work. Vivian Stanbury arrived in 1931, an acquaintance of the Camms from his childhood (the families were friends). Stanbury would go on to join the Project Office and eventually lead it after the war. Robert Lickley gained employment in October 1933 in the Stress Office, becoming much involved in the development of the Hurricane fighter, later being appointed to lead the Project Office in 1941.

Others joining the burgeoning design team before the war were Frederick Page in 1938 plus Frank Cross, Bob Copland, 'Digger' Fairey and Henry 'Roche' Rochefort. As the company began to grow at great pace with the various RAF expansion schemes generating a vastly increased factory output, the Hawker Drawing Office had come to be divided into experimental and production departments with experimental being responsible for new projects and drawings suitable for producing a prototype, and production responsible for taking those drawings once an order had been received and producing versions suitable for aircraft quantity production and the many and various modifications required once an aircraft was in service. Feeding information into the Experimental and Production DOs was the Stress Office, responsible for calculating loads and stresses on aircraft structures and approving the correct strength of materials to be used. Before the Second World War this staff included Leslie Appleton, who joined in 1936, Maurice Brennan, Joe Barrett and Charles Plantin. Another department was weights, under Chief Weights Engineer Ian Nightingale, which had responsibility for ensuring that structure was as light as possible for its given purpose and that weight distribution was correctly disposed around the aircraft centre of gravity (CG).

In the early stages of the Second World War, in 1940, following close calls from German bombing of Kingston (one bomb had hit the southern corner of the Hawker buildings, destroying the progress department); Claremont School, a mansion in extensive grounds near Esher, was leased for six years and the Design Office staff who dealt with new projects were moved out of Canbury Park Road and installed in the house. The production design staff, however, remained in Kingston. At this time also, Camm formalised the creation of a separate Project Office (PO), essentially a think-tank for the production of new ideas under the management of Robert Lickley. Their initial schemes, if sanctioned by Camm and the Hawker board, would flow from the staff in this office to the EDO for the creation of prototype drawings and thence to the experimental department on the factory floor to create a mock-up and later, if agreed, a prototype aircraft. At the same time, a numbering system was organised by Stanbury whereby each new project design was allocated a number, the first being P.1000 in 1940. Previously, identification was either by Operational Requirement (OR) specification number, e.g. F.3/44, or a simple description such as 'High Speed Monoplane'. The new system did not start at P.1 because Camm was insistent that he must have designed 'at least a thousand already'! The first staff members of the Project Office under Robert Lickley included Harold Tuffen, Vivian Stanbury and Ron Williams, who had started in 1943, Alan Lipfriend and Lesley Appleton (moving from the Stress Office and later responsible for armament integration) and Ken Bentley; the office reported directly to Camm rather than the EDO management.[4]

Ron Williams recalled his time at the Hawker Project Office in the 1940s. 'In 1943, when I entered Hawkers, the Kingston works had already been bombed

HG Hawker Engineering Ltd Design Office, Canbury Park Road, c.1927.

HG Hawker Engineering Ltd Design Office, Canbury Park Road, c.1927.

and the Project Office, along with most of the Design Office and services was based in the supposedly safe Claremont, Lord Clive of India's country mansion at Claygate, just outside Esher … The Project Office was located at the front on the ground floor, next to Stress. Robert Lickley was its head with Freddie Page (aerodynamics), Ken Bentley (structures), Alan Lipfriend (stability and control), Wally Walford (performance), and Vivian Stanbury (design layout). It did not get much bigger in later years, part of its success, perhaps.

'Although Sydney Camm was just across the corridor, I cannot remember Lickley ever allowing him into the Project Office. Lickley was a bit of a tyrant, hard on the senior staff but generous to us, younger mortals.

'In the Project Office … attention was being given to the possibilities offered by new powerplants to meet the current and envisaged military and civil requirements. There was even a tailless airliner project with boundary layer control air intakes along the swept wing…I suppose this incredible period ended in 1945, with the famous group photograph on the steps of Claremont.'[5]

Hawker design staff rose to their greatest numbers during the Second World War to cope with the huge demands for new projects with which to prosecute the fighting. Permanent staff were bolstered by those impressed into the company

Roland 'Roy' Chaplin OBE, Assistant Chief Designer, Camm's right-hand man and, latterly, Chief Designer himself.

under the Essential Work Order regulations introduced in 1941 by Ernest Bevin whereby the state could dictate one's employment choice and location. The numbers rose through the war from 146 in 1939 to 305 in 1945. While this was helpful to the company at the time, when restrictions were ended in 1946 the company lost a number of good design staff as they reacted to the relaxation of the regulations and left to seek employment of their own choice.

By 1946, with the war over, the Project Office and EDO were still resident at Claremont, not returning immediately to Canbury Park Road. However, with the relaxation of the EWO regulations, Hawker saw an exodus of talent from the Project Office. In 1946 Robert Lickley, head of the PO, left to take up the post of Professor of Aircraft Design at the new College of Aeronautical Engineering at Cranfield. He would leave there in 1951 to become Technical Director at Fairey Aviation, working on the Gannet and later the FD.2. He would then move back into the Hawker fold as a director of Hawker Siddeley Aviation. Replacing Lickley as head of the PO was Vivian Stanbury, Camm's family friend, who would subsequently leave in 1956 having received 'an offer he could not refuse' from Rolls-Royce Cars, to become their Chief Engineer. During his time in the PO, Hawker would move into the jet age and a rather more scientific outlook than had heretofore been required.

Claremont House, Esher, wartime home of the Experimental Design Office of Hawker Aircraft Ltd.

Freddy Page also left after the war to become Chief of the Stress Office at English Electric under W.E.W. Petter, working on the Canberra bomber and P.1 Lightning. Having gained a degree in mechanical science, he had joined the company in 1938, working in the Design Office. With the creation of British Aircraft Corporation (BAC), Page became a director, then Managing Director and eventually Chairman of the company, completing his career as Chairman and Chief Executive of British Aerospace and a knighthood to boot. Alan Lipfriend would also leave to become a barrister while Harold Tuffen moved back to the EDO as section leader, later becoming Head of Mechanical Systems.

Robert 'Bob' Marsh replaced Stanbury as PO head in 1956. Following his degree, he had found employment at RAE Farnborough, in the wind tunnels, before a move in 1941 to A&AEE Boscombe Down. His claim to fame was to have been one of the first passengers to experience jet flight when he was packed into the vacant ammunition bay of a Gloster Meteor to monitor some flight instrumentation around 1945. In 1946 he had obtained work at Hawker Langley with the 'Flight Test chap' and the following year, on the retirement of this 'chap', Marsh became head of Flight Test. In 1951, he moved to Kingston to head up Flight Development and was given a desk in the Project Office, becoming Head of the PO in 1956, retiring in 1981.[6]

Such then had been the wartime arrangements and personnel which had enabled Hawker to supply the RAF with capable designs, allowing the RAF to hold its own and then overawe the Axis forces ranged against the country. With the return of what was hoped would be a more lasting peace, at Kingston by 1948 the Project team was looking decidedly threadbare. But help was on the way in the shape of several exceptional engineers, young and very enthusiastic.

The first, interviewed that year by Camm himself, was Ralph Hooper. Hooper, on leaving school, had taken an apprenticeship at Blackburn Aircraft Ltd at Brough in January 1942, which included a stint at University College Hull before returning to Brough to work in the Stress Office. In 1946 he moved to Cranfield to undertake an aeronautics course under the Hawker alumnus Robert Lickley. Camm appears to have been impressed that a graduate engineer was seeking employment in his kingdom and soon had Hooper at work in the EDO before moving temporarily into the Stress Office. By 1953 he was in the Project Office. The other promising talent was one John Fozard who similarly obtained an apprenticeship with Blackburn before taking a course in aeronautical engineering and moving with a BSc to Cranfield to study, again under Lickley, gaining a diploma in aeronautics in 1950. On completion of his

Ralph Hooper OBE, latterly Divisional Technical Director, Kingston-Brough Division, British Aerospace.

studies, he joined the Hawker design team in the Stress Office before moving to the Project Office. With an honours degree under his belt, Robin Balmer joined Hawkers as a student apprentice in 1950, entering the Project Office in 1955 as a technical assistant, eventually heading the Project Office in 1963. Promotions in 1969 and 1971 followed with Balmer holding the positions of Assistant and then Chief Airframe Engineer, and Chief Project Engineer Harrier. Balmer was considered by all to be the stability and control expert so his arrival in the Project Office as the P.1127 and Harrier programmes progressed was well timed.[7]

Two other well-known names joined the PO in the 1950s. In 1956, Roy Braybrook arrived, in the final phase of a post-graduate apprenticeship (he had graduated in engineering from Manchester University in 1954) but was soon called up for National Service, returning in 1958 as a senior project engineer. His work would revolve around V/STOL studies and high-performance training aircraft. Roy later became a distinguished aviation author. Frank Mason arrived in late 1960; among his several jobs, he had been tasked by Camm with producing a company history, a monumental task which he finally completed. Button-holing Sir Sydney with the impressive results brought up to date, Camm's only response was 'So what? I assume you'll continue to keep it up to date.' (He did, the result running through several editions when published by Putnam.) The Project Office in the late 1940s and early 1950s, thought Hooper, was characterised 'by a burst of creativity in the future projects area as all the possible permutations and combinations of one, two or more jet engines with straight, swept or delta wings was examined'. Lastly, mention should be made of Trevor Jordan (performance), and Malcolm Ruscoe-Pond, Nigel Money and Arthur Green. Together, these engineers formed what Camm referred to as his 'young gentlemen'. Seldom more than a dozen at any one time, they were young and confident and held the future of Hawker Aircraft in their hands.[8]

At the top, however, at times it seemed that a firm understanding of the new disciplines inherent in design for the latest modern transonic aircraft was lacking. Camm, as head of design, should have been ensuring that his team's qualities fully reflected the requirements emanating from the Air Staff of the RAF. Yet it seemed that Camm's own design abilities had been left behind with the heyday of the biplane fighters which left him reluctant to admit his shortcomings, and tardy in ensuring that the Design Office was up to the job of producing for the air forces of the late 1950s and 1960s. Indeed, there was a feeling abroad that the Design Office had lost its way, that the sureness of touch, evident in the war had somehow been lost as labour moved away, firstly as a result of the relaxation of the EWO regulations and then later, after the loss of P.1121, to migration to the US and Canada which appeared to offer greater rewards. Notwithstanding concerns about the continuing effectiveness of the Design Department, in June 1953 Camm received a knighthood for services to aviation, becoming Sir Sydney Camm. T.O.M. Sopwith was also knighted.[9]

Hooper noted that, on his arrival in the Project Office in 1953, 'I wasn't all that impressed, we seemed to knock off pretty three-view drawings which frankly a schoolboy could have done. That's an exaggeration obviously but it didn't seem to me it was done on a strict enough basis really.' He felt that preliminary design could be done by any good draughtsman, given suitable guidance. Roy Braybrook

John Fozard OBE, latterly Divisional Director, Special Projects, Military Aircraft Division, British Aerospace Weybridge.

demurred, believing that such a person 'should also be able to do the preliminary weight and performance estimates … Preliminary design was not Ralph's forte … his forte was design development, which is a good forte to have, if you start out with a so-so design.' Fortunately, 'Ron Williams was the best preliminary design man, with an encyclopaedic knowledge of aircraft.'[10]

Brian Buss joined the Experimental Design Office in 1957 as a senior section leader, to work on the P.1121 project under Chief Draughtsman Frank Cross.

His interview left him underwhelmed and raised several questions: 'The result of that very poor interview was an offer for the post. I was somewhat perturbed by the questions it raised in my mind, e.g., were there no senior draughtmen in-house capable of tackling and leading the work? Was the company falling behind its own schedule? Did the company realise it only had a short window to prove this interceptor? Camm appeared to be beyond present aircraft technology. Am I stepping into the end of manned flight so should I stay put?'

Well, he did stay for a while and found himself immersed in a design environment that appeared to him to be something of a time warp.

'The structure of the Hawker set up was itself out of date for the period. It had a Project Office where ideas for new aircraft were born and when it decided to build a new design, it was passed to the Experimental DO for detail design and one or more prototypes were subsequently built by hand. If they proved to be acceptable the design was passed to the Production DO and the aircraft hopefully turned out in quantity. This was okay when aircraft were simple in design and could be easily constructed. However, they had become complicated and their many different systems required integration from the start. Also, few of the production and maintenance lessons learned could be understood or incorporated by the Experimental DO. This led to much bitterness between the two DOs, time wasting and delays. For these reasons I found the Experimental DO had had no use for a loft layout.

'The story was that the Production DO next door gave the Experimental DO an ultimatum at the start of the P.1121, "you either create a loft layout at Kingston, or it will establish its own to assist in the design and construction of major jigs and tools". As the design side would never permit the production side to dictate any part of its design it agreed, but to me it had no idea how to use it. The loft layout was based in a building once used by the film industry at Teddington and was managed by Tommy Wake who had no previous lofting experience. A pleasant individual but he did not have the personality to stand up to either Frank Cross or Henry [sic] Tuffen and no one in the DO appeared to exploit the service that he could offer design draughtsmen. How, I thought, had Hawkers managed to design aircraft like the beautiful Seahawk and Hunter without a lofting facility?

'I discovered that after the Company produced a hand-built prototype, production was contracted out to other companies who possessed a loft. For example, with the Hunter only the centre fuselage was produced at Kingston and this was completely circular, therefore no complicated double-curvature sections were involved. Some in the Production DO said the circular frames were laid out in chalk on the shop floor. I could not believe what I was hearing as companies like DH had, I understood, created a lofting facility way back. I know I wanted to gain experience in a large aircraft company, but not one some 15 years or more behind the rest.'[11]

Notwithstanding the somewhat archaic atmosphere in the Design Office, overseen and perhaps propagated by Camm, excellent products continued to flow from the fertile minds of the Kingston staff. Whether the Air Staff and politicians would similarly see them as excellent was another matter. Ultimately, under Camm's benign autocracy, the company would, by the time of his death, have produced 52 different aircraft types, constructing some 27,000 aircraft.

Although John Fozard's P.1121 project was slowly sinking (described later), in 1957 Ralph Hooper was toying with a small aircraft (also described later) built around a radically new engine project that offered the possibility of vertical take-off and landing – VTOL – and as work was run down on the P.1121, staff became available to assist Hooper in scheming a practical design around this engine.

In 1958 Hawker Aircraft obtained their first computer – a Ferranti Pegasus unit – approximately the size of a domestic sitting room. Although an analogue unit, it did point the way to the beginning of using computer programs to investigate some of the more arcane mathematical problems associated with aircraft design. Also installed was an Avro five-axis analogue computer, only the size of a table, to assist with aerodynamics, stability and control calculations. As time went by, computing capacity would grow and lead to more widely available access to this wonderful new technology but it would take some time and many of the older staff shied away from it whenever possible. Even Hooper admitted that he only ever did one computer program connected with reaction control bleed flow for the P.1127. 'After that I got more senior and there was always somebody else to do it so to this day, I haven't got a computer.'[12]

Of course, aircraft design is rather more than pretty lines on paper. Also requiring the utmost investigation are issues such as the stresses that the aircraft will experience during its life, the weight and durability of the materials used in its construction, the ability to carry out whatever function it has been designed to fulfil, not to mention concerns over stability and control and the aerodynamic qualities of the design.

The Stress Office was perhaps the most important adjunct of design for it was here that the choice of materials and overall design were tested to ensure that they did not fall apart in the sky and where fatigue testing was carried out on actual aircraft sub-assemblies to proof the airframe against failures caused by fatigue stress. That this was a possibility, even for Hawker, and was only too clear when the Hurricane replacement – the Typhoon fighter – was found to have a predilection for coming apart at high speed, the monocoque tail assembly parting company with the rest of the aircraft with inevitably fatal consequences. With all hands to the pump within the Stress Office, the staff as well as their work came under great stress, the calculations were checked and checked again with no fault being found. Eventually it was realised that the cause was, as recorded by Charles Plantin, 'remote concentrated mass balance on the wrong side of node in fuselage fundamental flex', leading to failure of the bracket securing the elevator mass balance, thereby inducing high-speed flutter and the consequential disastrous structural overloads causing failure of the fuselage at its weakest point – the transport joint between tail assembly and rear fuselage. This unacceptable failing in one of Camm's aircraft would cast its shadow for many years over the stressing of Hawker aircraft, which subsequently tended to be built like the proverbial 'brick ****house'.[13]

In the Stress Office (itself part of the R&D section) were Roger Dabbs, 'Nick' Nicholls and David Fowler under Henry 'Roche' Rochefort for many years. After the war, in 1947 Charles Plantin became Deputy Chief of Research and Development (Structures), responsible for running the test rigs at Kingston and Langley. The largest of the test rigs had been built at Langley, into which an entire aircraft could be placed and subjected to controlled loads simulating those encountered in flight. This, the Abbey Test Frame, was modelled on a similar structure at RAE Farnborough and provided much needed data on the loads that would be met as flight speeds and engine power increased. However, with the demise of Langley, new facilities would be required at Kingston. Colin Flint, one time Head of Ground Test Services, in recalling his time at Kingston, had this to say regarding the test frame: 'With the closure of Langley in the late '50s there was a requirement either to move the Abbey Test Frame to Kingston or build a new one so that structural testing could continue. Given that aircraft were getting larger it was thought sensible to build a new one. About the largest aircraft that Kingston could envisage building was something the size of a B-58 Hustler [a US supersonic bomber] and so that was the chosen size. People employed in the Research and Development Department were tasked with the design of such a test frame which was to be called Mithraeum after the Roman temple recently discovered during building work in London (all previous test frames were called after religious buildings). Derek Thomas was the lead engineer on the task.

'In 1959 the frame was constructed in Scotland and brought down by road in parts over a three-week period. Because of its height the frame was erected in a 10-foot-deep pit dug towards the northern end of the new Research building at the Hawker Aircraft site in Kingston [the Richmond Road factory]. This was cheaper than raising the roof of the 500-foot-long building by ten feet. The basis of the frame consisted of a pair of keel members some 95 feet long and overhead warren girders 105 feet long; these were mounted on four massive columns. Eight loading bridges were mounted on rails attached to the warren girders and associated structure, each having facilities for manual loading by turnbuckles via linkages to the structure under test. The pit was finished in waterproof cement so that it could be used for underwater pressure tests if required. A 23,000-gallon water tank and pumping facilities were included.

'The first major job was the P.1127 static strength test done in 1960, the load being applied manually via turnbuckles. Some twenty tests were carried out in various configurations, each taking about six weeks including rigging for the case, testing and data analysis. Each test required four people to apply the load, two to apply fuel-tank pressures (including myself), six to read the 800 strain gauges and four to read the deflection gauges. The maximum load required was divided into eight or ten increments, each of which was applied before the instrument readings were recorded. Each test took one day.

'As an aside, the last strength test carried out in the Mithraeum was the Hawk Mk 1 test series where the six weeks per case was reduced to half a day using automatic load measurement and mechanical servo-controlled hydraulic valves. Just four people ran the test and graphs of strain gauge and deflection measurements were available one hour after the test was completed. The next major test was a research programme on twenty Hunter Mk 5 airframes for the RAE (Farnborough). The RAE was to test using constant load cycles from start to finish at various levels; Kingston was to use programmed loading (a programme consisted of 311 cycles). Two airframes could be installed in the test frame at the same time. On completion of each simulated 500 flying hours inspections were carried out overnight.

'The next test carried out in the Mithraeum was the Harrier GR. Mk 1 strength test. The wing failed marginally below the required load and as a result the wing skin thickness was increased on all service wings. These tests were followed by the Harrier GR. Mk 1 fatigue test and then the T. Mk 2 fatigue test.

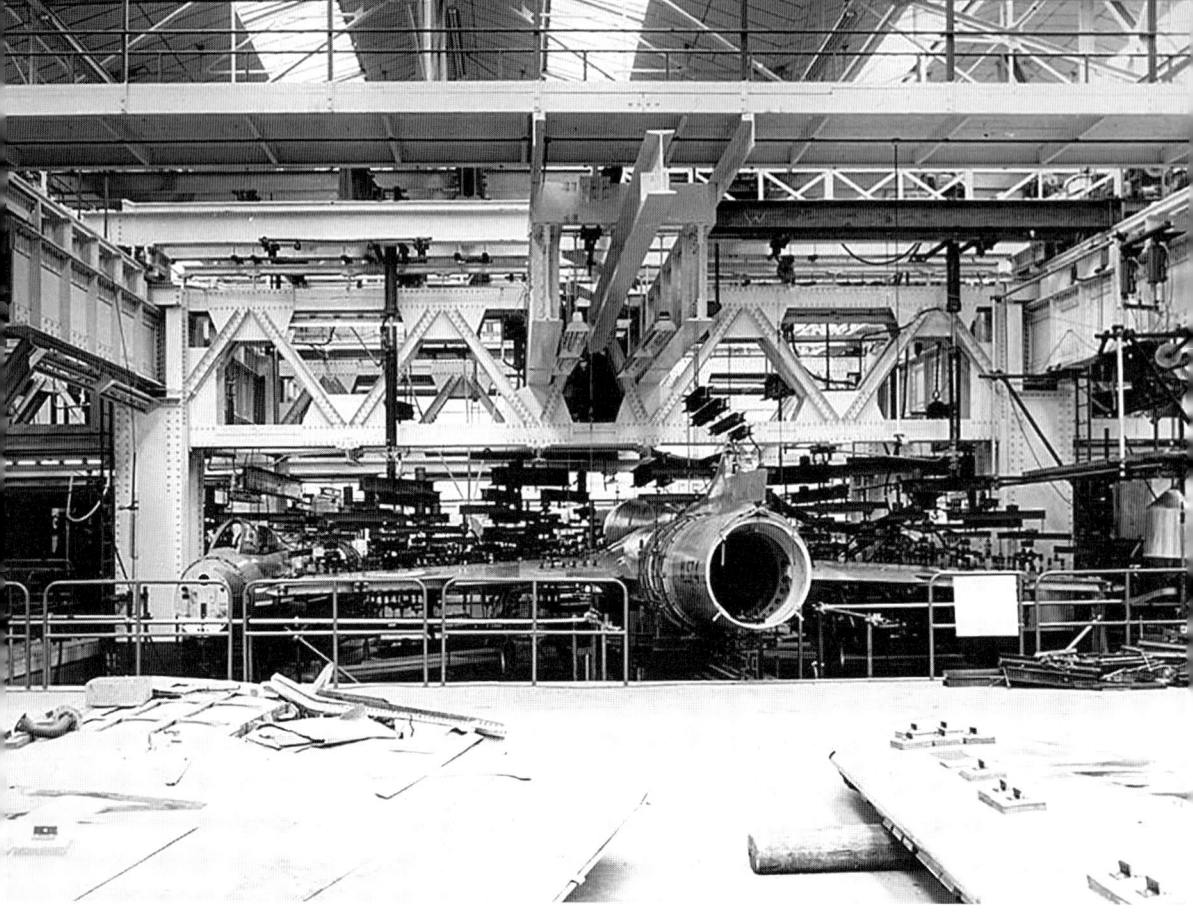

Hunter aircraft under load test in the Mithreum, Hawker Aircraft Ltd, Richmond Road, Kingston upon Thames.

Both used the by now well tested system used on the GR. Mk 1 strength test but the mechanical hydraulic servo valves were motorised so that case changes could be accomplished without human intervention during testing. Both these tests reached 200% of the required life.

'The final full-scale tests carried out in the Mithraeum were the Hawk strength test, mentioned earlier, followed by the first Hawk fatigue test which was stopped at 60% of the required number of cycles when the wing failed, there being too much damage to make repair an option. This test was the first to apply simulated manoeuvre loads, for example rolling pull-outs and Cuban eights. Specific sortie patterns were also run.'[14]

Whilst the Stress Office appeared to be well catered for, the same could not be said for aerodynamics. It was perhaps unfortunate that this was one of several areas which represented for Camm something of a bête noire. It was a matter of personal honour for him that he did not need an aerodynamics department because 'when you've designed aircraft as long as I have, you can see the airflow!' (on the subject of area ruling of the Hunter by RAE in 1956). All well and good in the early biplane days but, as design moved more and more away from Camm's 'stick and string' comfort zone, so it became more of a problem and a drag on modern design within the department. As a result of Camm's stance, Kingston never came to possess a wind tunnel, a fairly fundamental requirement when designing for the future. Instead, the Design Office relied on use of the wind tunnels at RAE Farnborough (where they were glibly referred to as the 'Hawker' tunnels), the National Physical Laboratory (NPL) and, from 1956, the Aircraft Research Association (ARA) wind tunnels at RAE Bedford. Camm's bias against wind tunnels meant that when the NPL recommended a t/c ratio of 19 per cent for the Hurricane wing, Hawker was not in any position to check the results for themselves.

As it happened, for the Hurricane, the wing thickness played in its favour in allowing easy accommodation for the guns whereas the Spitfire proved to be rather more complex though, in later life, Camm appreciated that with a thinner wing the Hurricane would have been an even more effective fighter. When the Hurricane's successor, the Typhoon, was subjected to wind-tunnel testing by NPL, the recommendation was again for a thick wing with the result that, as a fighter above 20,000 feet, the aircraft was out-paced by the opposition. A thinner wing, as belatedly appeared on the Tempest, would have resulted in the Typhoon being an outstanding fighter at high altitude. However, come what may, somehow or other, Hawker designs were models of aerodynamic efficiency whenever possible which, considering designs such as the P.1127, wasn't always easy to achieve.

The Weights Department or, as Camm would have it, the 'wait' department, was principally concerned with the use of materials in the design which would give the greatest strength for the lowest weight. Also considered there would be the distribution of weight through the airframe and the position of the CG. Control and stability were vital concerns for the design process. Although stability was considered a desirable aspect of early design, for a modern fighter, the ability to manoeuvre is of paramount importance during attack and defence and, for this, instability is often the key. Today, most fighter aircraft are designed to be substantially or completely unstable, control being maintained via computerised control. Further consideration for the designer is adequate control of the aircraft, and in its basic form this requires that the control surfaces are as efficient as possible and not influenced by other aspects of the design. This was shown to be vital during the late 1950s and 1960s when the high-mounted 'T-tail' was much in vogue. This kept the tailplane and elevators clear of the wing and engine disturbance very nicely but it was found that, in a stall, the wing could blanket the elevators making it impossible to recover and lead to a catastrophic crash.

As well as the Stress Office, the other department responsible for feeding information back into the design function was Flight Development. At Langley this comprised of one man who collected and analysed the data from test flights and passed it back to Kingston. When, after the war, Bob Marsh joined him, the department expanded by 100 per cent. With the advent of the early Hawker jets, the importance of Flight Test increased considerably, as did the staffing and, firstly, at Farnborough and then at Dunsfold under 'Fred' Sutton, the work and responsibilities of the office increased significantly, forming a leading part of the processes carried out at Hawker's flight test centre at Dunsfold.

Lastly, what is now called systems engineering came to the fore during the 1960s. This was concerned with approaching the areas of aircraft hydraulics, electrics, navigation, engine control etc., not as separate entities but as unified areas of concern, to be designed into the aircraft from the initial proposal.

Allied to this unified approach was that called weapon systems, again an integrated area of investigation which considered the aircraft's weapons not as some random piece of equipment fitted as an afterthought to the aircraft but designed in from the start. The designer could no longer sit back on their laurels, having produced a beautiful flying machine; its function must be considered right from the outset, the aircraft almost then becoming an afterthought, emerging from the need to produce the best vehicle with which the weapon could be utilised. It was in the later 1950s that the weapon-system concept started to come to the attention of Hawker designers, courtesy of the Americans, of course. In its widest sense, this saw the aircraft and its weapon as part of an integrated air-defence system comprising ground-based early-warning radar, manned missile-armed air interceptors and ground-to-air missiles. In its use as applied to aircraft, it related to the weapon, target acquisition and on-board weapon control as an integrated whole.

Sydney Camm, never one to take on radical new concepts unquestioningly, was well aware of the calls emanating from Whitehall, via the OR branch of the Air Ministry, that future designs must reflect the weapon-system concept. Having suffered heavy criticism during the P.1121 period that the design did not have sufficient weapon-system capability, Camm realised that he would have to respond to the demands in some fashion. Initially, Bob Marsh had offered the role to Roy Braybrook who 'turned it down for obvious reasons'. It fell to Ralph Hooper to be singled out to take on the responsibility for the process, much against his better judgement. At this point Camm was content that he had sorted the issue, off-loading the problem onto Hooper and completely missing the point, probably knowingly, that weapon-systems integration was a philosophy intended to permeate the entire design process, rather than an add-on to be applied later. At one point Camm entered the Project Office and showed Hooper a letter to the Air Ministry stating in part, 'We have now set up a weapons systems cell'. Laughing, he pointed at Hooper and said, 'That's you!' Luckily for Hooper, work gaining pace on the P.1127 design, for which he was for a while solely responsible, saved the day and he was able to quietly drop his weapons-systems epithet. Eventually the role was picked up by a team from Brough.[15]

As the 1960s gave way to the 1970s and 1980s, electronic aids began to permeate the designers trade, firstly as vast analogue computers, later replaced by digital units, and then as desk-top equipment, electronic calculators and, eventually, computer terminals, sealing the fate of the beloved slide rule for complex calculations – the digital office had arrived!

From 1957, Hooper had first begun to toy with what might possibly be designed around Bristol's strange new engine proposal. The eventual result would go on to become one of the main focuses of work in the Design Office through its various iterations as the P.1127, Kestrel, P.1154 and, ultimately, the Harrier. Hooper had led on the V/STOL concept, it being 'his' design, while Fozard had taken on the naval aspect of Hooper's P.1154, becoming Chief Designer P.1154, and later the Sea Harrier design lead. Roy Braybrook was not particularly impressed with Hooper's original wing design.

'Ron [Williams] and I told him that the initial wing design would be a disaster, but he insisted that "If Douglas [Aircraft Company] can get away with it so can we". In reality the A-4 [Skyhawk naval fighter] design, regarded by some (not me) as a miracle, had a light wing that achieved a good low-speed co-efficient thanks to its slats. It didn't have the problem of [outrigger undercarriage] wingtip pods, and its high Mach lift co-efficient was nothing remarkable. Six designs later, the Harrier had a really outstanding wing.'[16]

Eventually Hooper became Chief Engineer at Kingston while Fozard became Chief Designer P.1127 (RAF) i.e., Harrier. The various Harrier marks and the work to achieve a next generation STOVL fighter would continue, unsuccessfully, for the duration of Hawker and Hawker Siddeley's existence, and beyond. Ultimately Ralph Hooper would retire as Divisional Technical Director, Kingston-Brough Division, British Aerospace. John Fozard (whom Braybrook considered 'the only one of the trio [Hooper, Fozard, Williams] equipped with the indestructible constitution needed for long-term survival in the role of a modern chief designer') would retire as Divisional Director of Special Projects, Military Aircraft Division, British Aerospace.[17]

In the early 1960s, the various companies gathered under the umbrella of the Hawker Siddeley Group were re-organised into two separate entities, Hawker Siddeley Aviation and Hawker Siddeley Dynamics. As seen above, into Hawker Siddeley Aviation went Avro, Blackburn, Gloster, Armstrong Whitworth, Folland, de Havilland aircraft companies, as well as Hawker Aircraft at Kingston and Dunsfold. This new grouping formalised the loose collection of companies previously trading within the Hawker Siddeley Group. From July 1963 Kingston and Dunsfold were formed into a division with the old Blackburn factory at Brough to form the Hawker Blackburn Division. This closer working relationship resulted in some movement of design staff between the sites, mainly of Brough staff to new lives in Kingston. Also migrating to Kingston and Dunsfold were former members of the Folland design team from Hamble and Chilbolton, who would later be involved in the design work for the project that resulted in the Hawk trainer. Other changes were in play at this time. In 1959 Camm was persuaded by the Hawker Siddeley board to take a new title: formerly Hawker Chief Designer, he became Chief Engineer of Hawker Siddeley Aviation and a director on the HSA board. Roy Braybrook noted that 'Camm stuck to the CD title like glue. It was an office joke!' Nonetheless, this 'sideways' promotion cleared the way for his deputy, Roy Chaplin, to take over as Chief Designer though little changed in Camm's overview of the Kingston design function. In the event, Chaplin would not hold this position for long, a heart attack prompting early retirement in 1962.

Included in this migration were the two 'Gordons', Gordon Hodson and Gordon Hudson, (a recipe for constant confusion!) both subsequently influential in the Hawk trainer programme. Hodson had cut his teeth on de Havilland products at Christchurch before joining Folland in 1956 as a flight test engineer on the Gnat fighter and, later, trainer. Moving to Kingston in 1965 he continued on Gnat systems before becoming Head of Preliminary Design for what became the Hawk trainer and then Assistant Chief Designer Hawk. In 1986 he was responsible as Executive Director for the T-45 Goshawk programme for the US Navy, retiring from BAe as Project Director T-45A Goshawk in 1991. Hudson had joined the Folland team towards the end of the war and graduated in 1952 to Chief Stress Man on the Gnat fighter. Involved with major components for Hawker's V/STOL projects from 1957 to 1965 at Hamble, with cancellation of the P.1154, he moved up to Kingston, becoming Assistant Chief Designer (Airframe) and, in 1971, Chief Designer Hawk and eventually Executive Director and Chief Engineer (Kingston).

Around 1980 the organisation of the design facility at Kingston underwent significant changes. The Project Office function was split into two new departments. The Project Office had been renamed Future Projects in 1969 and this was now split off from the other disciplines which were incorporated into a new enlarged department entitled Airframe Engineering which covered the aerodynamic specialities (performance, stability and control) together with stress, structural dynamics and weights. The various offices then comprised Future Projects, Airframe Engineering (Aerodynamics and Structures), Systems Engineering (Mechanical and Avionics), Test Engineering (Flight and Ground Test Services), Drawing Office (Experimental and Production) and Engineering Management and this structure would broadly endure until closure in 1992.

Design staff numbers throughout the period from the 1920s up to the end of the 1950s tended to reflect the success or otherwise of Hawker's products. From thirty-one staff in 1925, this had risen to forty-six at the advent of the 1930s. A steady increase was then apparent up to 1939 (146 staff) to war's end in early 1946 (321 staff). Thereafter the numbers dropped away as staff left for other pastures (210 in 1949), before climbing again as Hunter and P.1121 work came on stream resulting in a steady increase to 396 staff in 1957. At this point, in 1957, with the cancellation of P.1121, numbers dropped off before rising again to nearly 550 to cope with P.1127 and P.1154 before another drop with the cancellation of the latter project. A slow rise then saw staff numbers increase to around 620, only to drop back again with the imposition of the government's pay policy restrictions. With the return of a Conservative government under Margaret Thatcher, numbers then rose again to nearly 800 in 1985. At this point, at the end of 1985, Kingston's design offices were moved to Weybridge, then the headquarters of BAe, and merged with the Weybridge Future Projects team. This was possibly a precursor to plans to close Kingston site completely. However, given the heavy workload at Kingston and the concomitant lack of work at Weybridge, it would be the latter which would close first, in 1987, and therefore Future Projects returned to the Kingston site.[18]

Sadly, in 1988, under Chris Hansford, the Future Projects Office at Kingston upon Thames succumbed to the internal politics of British Aerospace plc. Having suffered the indignity of wholesale relocation to Weybridge, effectively cutting them off from the factory and offices at Kingston, in 1988 the office

Kingston 'unofficial' artwork illustrating the perceived battle between Warton (with EFA) and Kingston (with Harrier) design groups.

was closed completely. Henceforth, Warton would champion future BAe and BAE Systems designs.

This then is the background to the projects described in this work, none of which ever flew, at least not in service. In that sense, perhaps this should be seen as the other side of the many successful designs to flow from Hawker and Hawker Siddeley. The reasons behind the failure of the projects are varied but ultimately derived from the decisions of the government of the day not to proceed with orders. Whether this stemmed from the project designs not fulfilling the specifications emanating from the Operational Requirements branch of the Air Ministry or flaws in the basic design will become clear as each design is discussed in detail later.

Chapter 3

Hunter Developments

In concluding their report on 'Science at War' in 1945, the authors had this to say regarding radar. 'Radar, in consequence, has played a great and increasing role right from the beginning of the present war. It has, more than any other single development since the airplane [sic], changed the face of warfare; for one of the greatest weapons in any war is surprise, and surprise is usually achieved by concealment in the minutes and hours before an attack … [which] simply doesn't exist in the world of radar. The tactical thinking of an attacker or a defender must take this fact into account.'[1]

The pre-war development of radar in the UK originated from an urgent requirement to find some means of early warning of approaching enemy bombers and this was achieved in time – just – for the Battle of Britain in 1940. But while the Chain Home (CH) radar network and the tracking and reporting system into which its raw material was fed offered good warning and protection during daylight, in the hours of darkness the country was effectively defenceless. To address this problem, Dr E.G. Bowen's team at Bawdsey under Robert Watson Watt had, in 1936, begun to investigate and develop a radar system small enough to be carried aloft by an aircraft, and with it to locate and destroy enemy intruders. By 1940 these efforts were slowly moving towards a workable system with a much shorter wavelength than the cumbersome CH system but an operational airborne interception (AI) system would come too late to have any effect on the night-time Blitz that devastated London and many other cities.

However, by the end of the war, sufficient progress and experience with AI had seen the static antennae replaced with a small and compact elliptical dish array suitable for mounting in the nose of a fighter which allowed a target to be not only illuminated by the radar but also had the ability to lock-on to the target and track it automatically. With the advent of early guided missiles and the ability to carry them on aircraft, the two concepts were engineered to work in concert so that a radar beam could illuminate the target allowing a missile to be fired which would follow the beam until impact with the target. As miniaturisation reduced the weight and size of the equipment, so the size of the carrier aircraft could also be reduced.

The Missile Armed Hunter

The first 'beam-riding' missile, code named Blue Sky, was developed by Fairey in 1949, weighed 330lb (50kg) and was about 9.3 feet (2.83 metres) long with a speed of Mach 2. Under the name Fireflash, collaboration with Hawker Aircraft saw the missile applied to the Hunter for carriage trials, an F.4 aircraft, XF310, being allocated to Fairey in July 1956 for trials. Still under project code P.1067, the initial scheme showed carriage of four underwing missiles, though the flight trials only involved two missiles carried on the inboard pylon position. Fireflash was an unpowered missile, propulsion being obtained from strap-on rocket packs which were jettisoned after initial acceleration to around Mach 2. While it was effective in good conditions, it relied on the pilot keeping the beam (generated by a modified radar-ranging set) on the target until impact, not necessarily possible in a combat situation. Poor results led to the system being abandoned, though it was resurrected later in heavily modified guise by other manufacturers rather more successfully.

This technology was soon overtaken by infra-red guided missiles which, once the aircraft radar had located the target, could be fired and left to seek the enemy without further pilot intervention, the missile's infra-red seeker locking on to the heat from the opponent's jet engine. Under the code name Blue Jay, de Havilland Propellers developed this technology and marketed it as Firestreak. Blue Jay weighed 300lb (136kg), was 10.5 feet (3.2 metres) long and had a closing speed of Mach 3. Again, Hawker Aircraft, following encouragement from the Guided Weapons Division of the Ministry of Supply, collaborated with the manufacturer, equipping Hunters to carry the missile (under specification F.167D).

Initial carriage trials were completed using WB202, the third prototype, in July 1954 before Hunter WW594, under Hawker's designation P.1109A, was modified with a new radar-capable nose for more in-depth trials, with two of the guns deleted to maintain the CG. This aircraft had the radar but not the missiles. It was followed by XF378, under designation P.1109B, which was fully equipped with AI.20 radar, code named Green Willow and Firestreak missiles for detailed firing trials, carried out at RAF Valley. Green Willow was an x-band radar developed by EKCO Electronics which gave good results but was ultimately cancelled in favour of AI.23 which would go on to equip the P.1B Lightning interceptor. Blue Jay would be upgraded as Blue Vesta before emerging operationally as Red Top, equipping the Lightning fleet. The complete Firestreak missile installation, not including any radar, was surprisingly heavy, the complete four-missile fit weighing in at 1,539lb, pushing the CG some 2.6 feet aft, though an alternative fit of two missiles on the ventral gunpack was also drawn which would have been somewhat lighter.

The trials with guided missiles wedded to the Hunter showed that such a combination was certainly feasible and, particularly with AI.20 (or AI.23) and Blue Jay, offered a relatively inexpensive means of putting a radar-guided weapon system into operation but the contract was cancelled in May 1956. However, a letter from Mr Serby at DGGW was received, generously encouraging Hawker to continue with research at their own expense. In the event, the company did elect to continue trials with one aircraft from which two missiles were successfully launched.[2]

In cancelling the missile Hunter trials, it appeared that the Air Ministry had other plans, founded upon large all-weather fighters such as the Gloster Javelin, de Havilland DH.110 Vixen and the single-seat English Electric P.1B Lightning. Indeed, such was the concern at the Ministry that Hawker might steal a march on their plans that the company was forbidden to demonstrate the missile-equipped Hunter at Farnborough wearing RAF insignia lest anyone get the frightful idea that it had any support within government. Later modifications by some Hunter export customers would see the aircraft go into service equipped with the Sidewinder missile, so eventually the Hunter did become operational as a missile carrier.

This would not be the last time that Hawker's inexpensive answers to the RAF's expensive requirements for shiny new aircraft got up their collective lordships' noses within the Air Staff. Although the all-weather DH Vixen did not enter service with the RAF, if Hawker's missile Hunter was seen to be entirely capable of performing the task that the forthcoming Lightning was designed around, then the amazing Mach 2 interceptor might not be needed at all. This irritating propensity of Hawker Aircraft Ltd to interfere with the plans of the Air Ministry would surface again during the debacle that was TSR.2.

The Hunters purchased by Sweden in 1955-56 were modified in-country to carry the Philco Sidewinder infra-red missile under the wing, greatly increasing their combat potential. The carriage of guided missiles on the Hunter received a further fillip from David Lockspeiser, one of Dunsfold's test pilots, when, in 1962, he schemed a proposal to carry two AIM-9 Sidewinder missiles on the Hunter gunpack. This scheme involved the removal of two of the four cannon from the pack which gave space for the missile electronics and retained the correct CG. Although received favourably by the RAF, the Air Ministry was

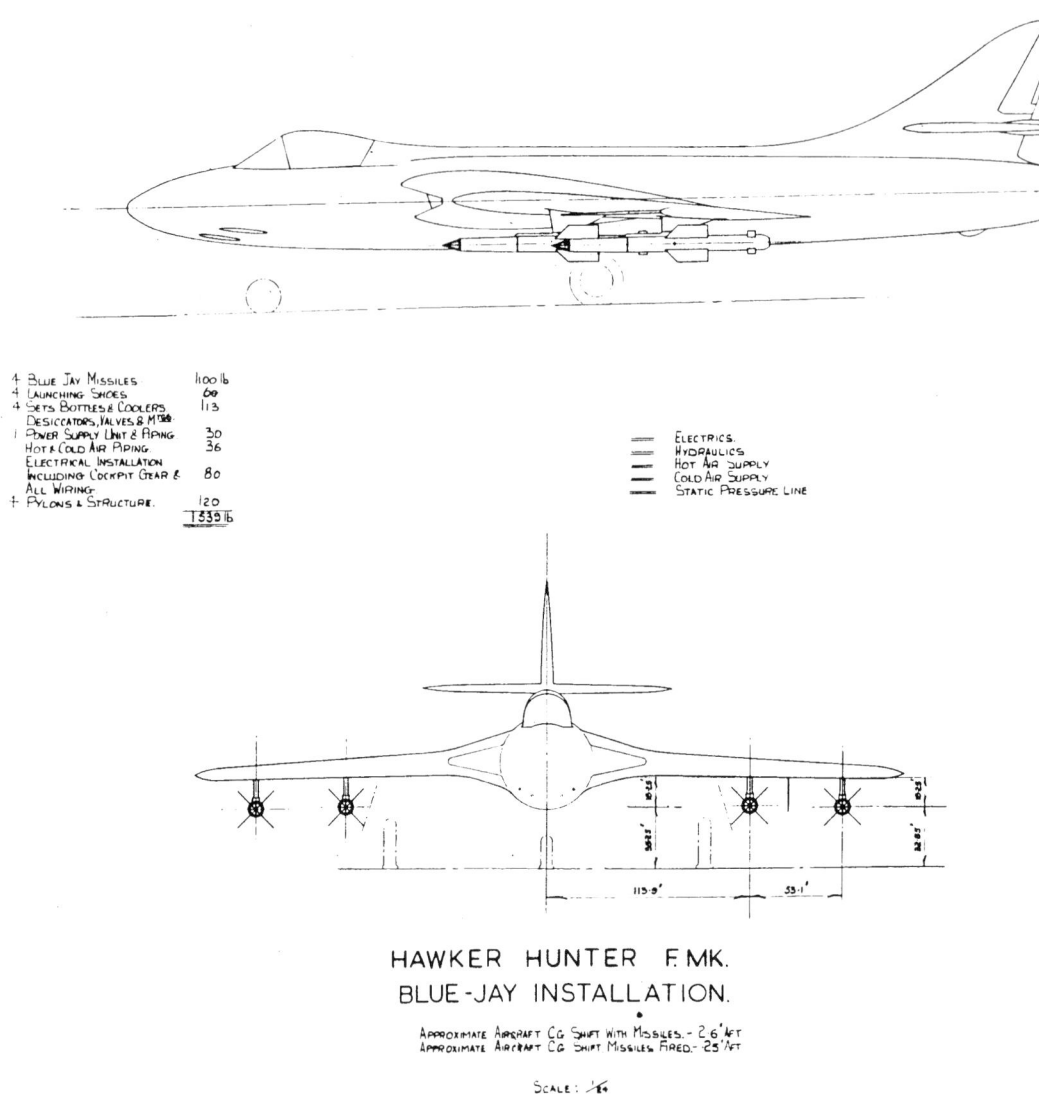

Hunter F.4 GA with 4 x Blue Jay missile installation.

less enthusiastic, claiming that the Hunter would be retired from service in a few years anyway. However, missile-armed Hunters did enter service with the Republic of Singapore Air Force and Switzerland as well as Sweden, though the missiles were carried on wing hardpoints. Also drawn at this time were alternative schemes for carriage of Nord AS.30 and Bullpup air-to-surface missiles on the Hunter FGA.9, as well as Firestreak carried on the gunpack location.[3]

Hunter Developments 51

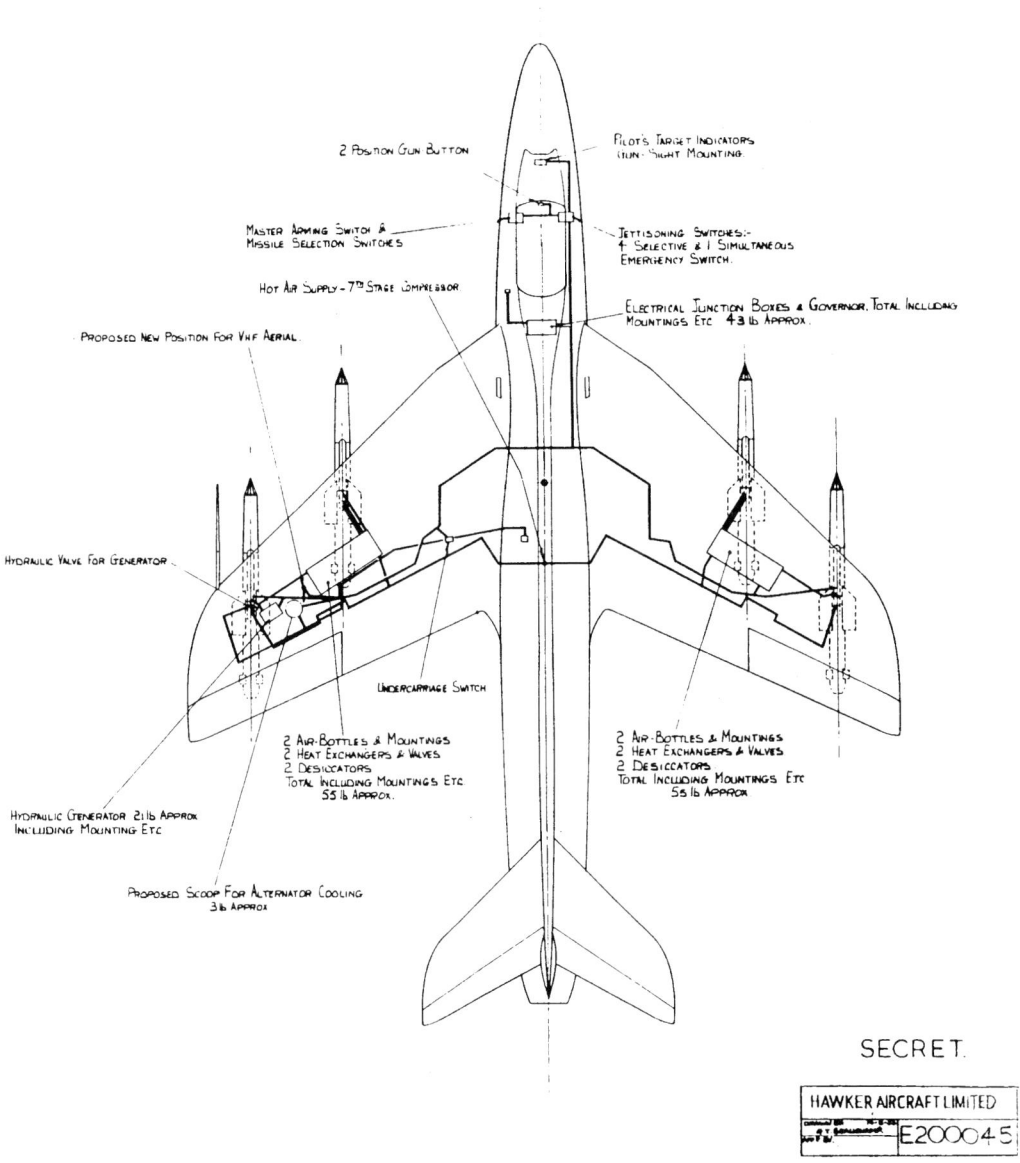

The Hawker Hunter, one of the UK's most successful military aircraft programmes, had been conceived in the early days of the Cold War. No sooner had the threat from Nazi Germany been finally overcome than the new threat from a resurgent USSR came to dominate western European defence doctrine. Specifically for the UK was the perceived threat of fast, high-flying bombers, sweeping in across the North Sea upon an exposed south-east of England. While radar would give some warning of the impending arrival, it was seen as crucial

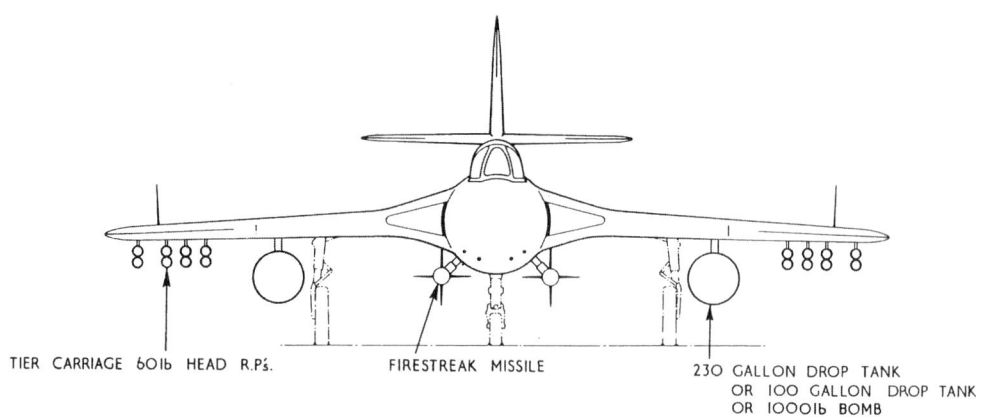

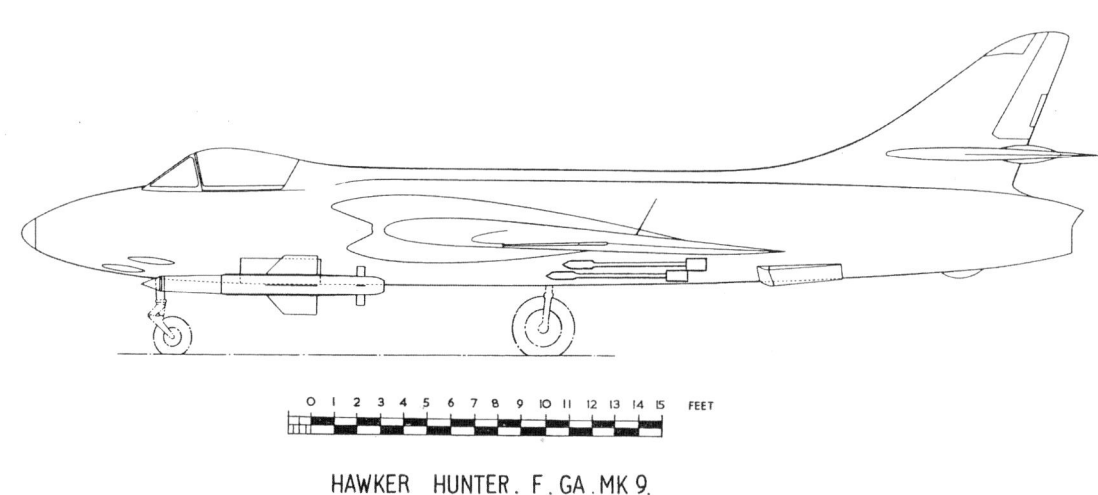

Hunter FGA.9 GA with 2 x Firestreak missile installation on gun pack.

that fighters engage the advancing armada as far out to sea as possible if they were to have any chance of destroying sufficient targets to deflect the impact of the attack. To this end, the RAF and Air Ministry sought fighters capable of climbing as quickly as possible from their bases to the altitudes at which the bomber stream was expected to fly and armed with a sufficient lethality of weaponry to enable the fighters to destroy their target in one quick pass.

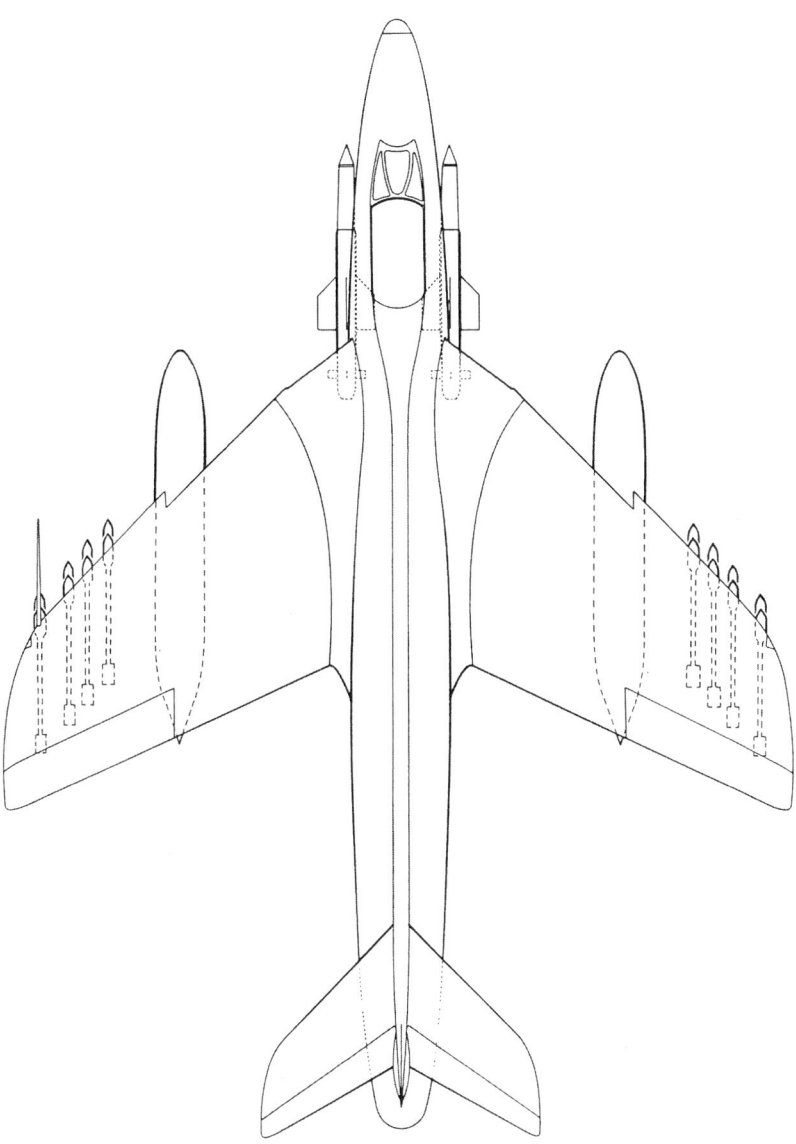

With the acquisition of nuclear weapons, the threat from the USSR increased exponentially. It was depressingly realised that, while attrition of enemy bomber formations during the recent war had been considered good if 3 per cent had been brought down, nuclear-armed bombers would require something in the region of a 90 per cent attrition to be in any way considered a successful defence. Into

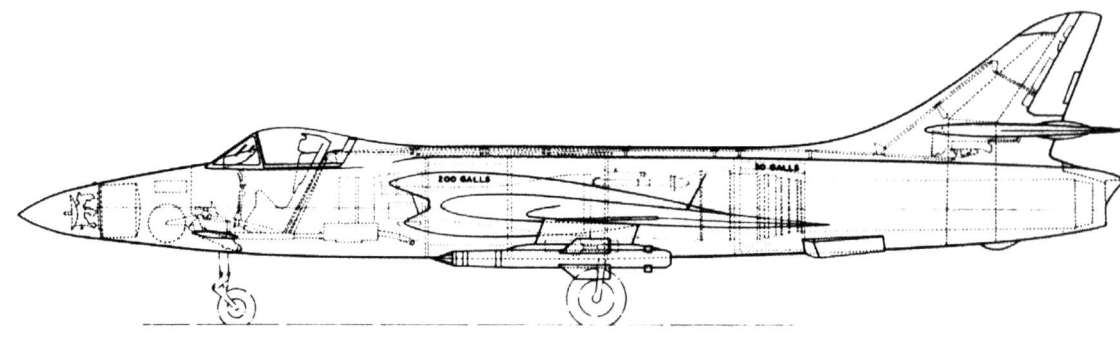

SPAN: 33 FT. 8 INS. WING AREA GROSS: 349 SQ FT.
O/A LENGTH: 49 FT 1 IN. ANGLE OF SWEEPBACK: 40° (¼ CHORD)
FUEL CAPACITY: 390 GALLONS TWO 30mm ADEN GUNS 150 ROUNDS EACH.

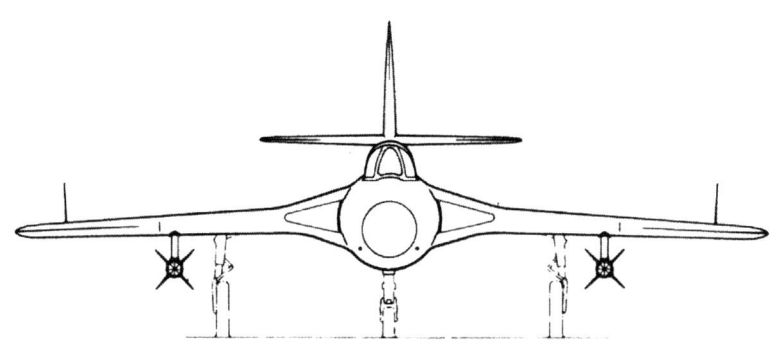

HAWKER P.1133 ALL-WEATHER FIGHTER
SINGLE-SEAT HUNTER F.6 WITH A.I. 23 RADAR AND FIRESTREAK
ROLLS-ROYCE AVON Mk. 203 TURBOJET

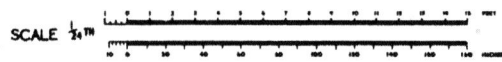

Hunter F.6 GA with 2 x Firestreak missile and AI.23 radar installation.

this sobering environment, the Royal Air Force, its Ministry and UK aircraft companies set to, to consider just what form of defence was possible.

The initial response had been an operational requirement, OR.228, and two specifications, F.43/46 and F.44/46 for a day fighter and night/all-weather fighter respectively, the F.43/46 specification being aligned fairly closely to Hawker's

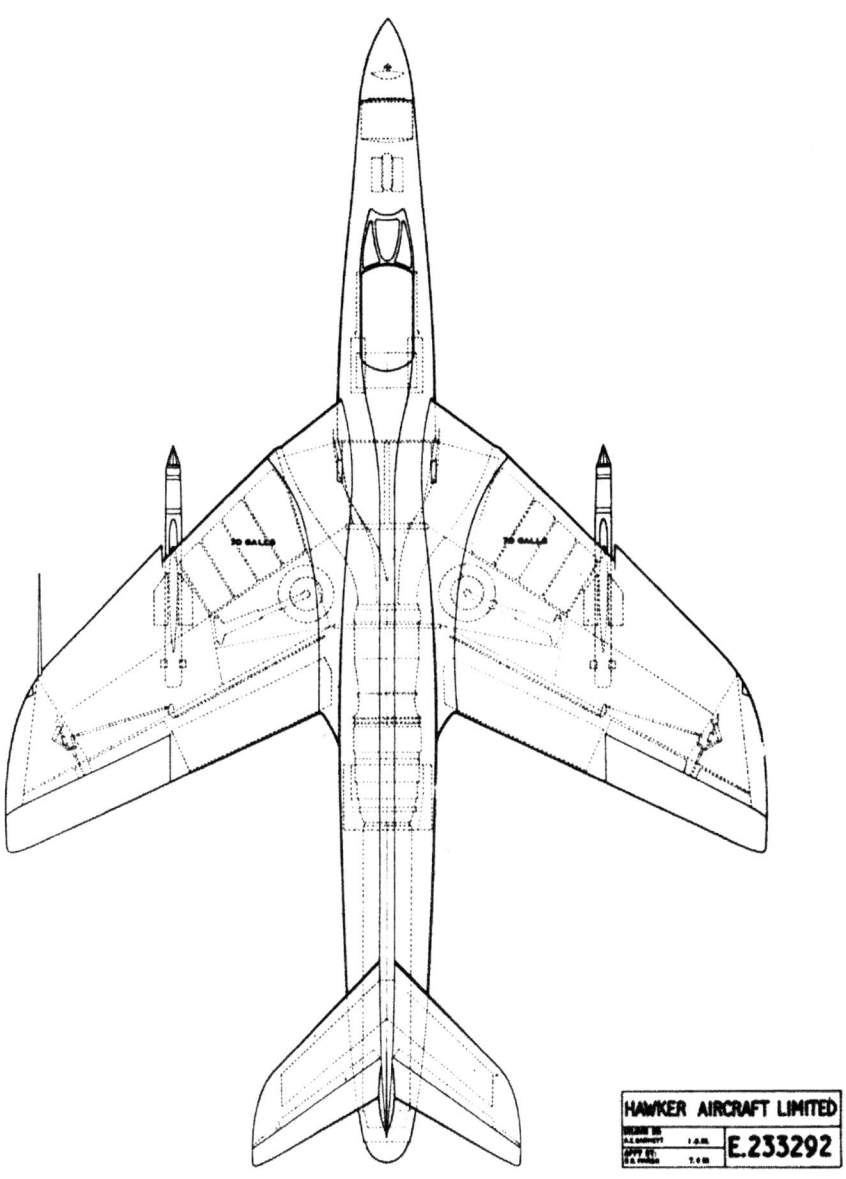

P.1054 project (a twin-engined design). Although Hawker had subsequently responded to the invitation to tender, neither Camm nor the Air Ministry believed that this was the definitive answer to the requirement. Having spent most of 1947 attempting to produce a successful design to the specification, Sydney Camm admitted defeat and instructed his project engineers to look to a slim single-engined, swept-wing design instead.

Hunter P.1099A, WW594 with radar nose and 2 x Aden cannon at Dunsfold.

Work began in January 1948 on an aircraft of some 12,000lb weight armed with two 30mm guns, this initial approach being accepted by the Ministry which requested that the company continue on this design. Advice from RAE was 'to keep the project as slim and symmetrical as possible. The 8.5% thickness/chord mid-wing and 50 inch diameter fuselage closely wrapped round the Avon engine was the result.' (Initial drawings show thickness/chord ratio as 9.5 per cent and sweep back at 42.5 degrees). Hawker investigated at least twelve alternative layouts using four draughtsmen, involving wing loading, exhaust reheat and rocket assistance together with single and twin-engined configuration before a revision of OR.228 led to a new specification, F.3/48, which was issued to replace F.43/46.[4]

From this uncertain beginning, the prototype P.1067 Hunter would make its first flight from A&AEE Boscombe Down in July 1951 and was hurried into service in 1954. Despite its early vicissitudes, the Hunter would go on to equip the majority of the UK's day fighter squadrons until replaced by the English Electric P.1B Lightning, the Hunter later becoming a very effective ground-attack aircraft both for the UK and for numerous other air arms around the world, remaining in service until the end of the century and in some cases beyond.

It was axiomatic in the aircraft industry that a winning design would be the subject of further development in an attempt to wring more success from an already successful product. For Hawker Aircraft Ltd, this had proved a winning formula with the Hart series of biplanes in the 1930s and the Hurricane during the war. Now, with Hunter sales expected to be significant, the company naturally set about the job of attempting to gain more performance from their transonic design. Even by the early 1950s, the pursuit of supersonic speed was engulfing Air Ministry minds, sometimes to the detriment of rather more prosaic requirements. The advent of the English Electric Lightning was perhaps the ultimate result of this philosophy – very fast but notoriously short of fuel and a weapon system of dubious reliability.

The Supersonic Hunter

It was perhaps natural then that Hawker should seek to gain supersonic speed from its Hunter airframe and this it proceeded to do with the assistance of Rolls-Royce. Even before the F.3/48 Hunter prototype had flown in August 1950, the Directorate of Operational Requirements within the Air Ministry (DOR/C) had carried out an assessment of the likely performance of the P.1067 aircraft (or F.3, the short form of F.3/48, as it was termed within DOR), particularly when flown against the new Soviet fighters. In a note to the Ministry of Supply, DOR stated that 'although the F.3 was superior to the MiG-15, the margin was not outstanding. The MiG-15 was already in service and presumably was capable of some improvement. The Air Staff were therefore interested in any plans the Ministry of Supply might have for developing the F.3.'[5]

As it happened, the Air Ministry was knocking at an open door. Even as the design that would become the Hunter was in its earliest stages, ideas for its development were being discussed in Kingston's Project Office. In September 1948, the Hawker investigation into a transonic design bearing the project code P.1069 was forwarded to PDSR. This retained features of F.3/48 as it then stood, with nose air intake and tailplane mounted low on the fin, but with 50-degree sweep on the wings and alternative engines, Avon or Sapphire, both with 20 per cent reheat. This was followed the next month by P.1071, a similar P.1067 design with reheated Avon but with the addition of a 2,000lb thrust rocket in the tail. A preliminary visit by DOR to Hawker in May 1950 had resulted in initial discussions on such a project which would feature a wing sweep of 50

Hunter x 4 aloft showing (nearest the camera) XF310 with Fireflash missiles, XF378 with Firestreak missiles, Hunter with multiple rocket fit and a further Hunter with 4 x drop-tanks.

degrees to facilitate supersonic speeds in level flight, with a note being written up the following month on the project, now coded P.1083.

Following discussions with RAE on 31 May 1951, a technical brochure was submitted by the company to DMARD showing a development of the F.3/48 design with 50-degree swept wing and AS Sapphire engine (some drawings by Fozard show two AS Snarler rockets mounted under the tail at this time). In November 1951 serious work in the design and stress offices at Hawker Aircraft on the P.1083 began while the following month, on 10 December, at a conference at the Air Ministry it was agreed that the proposed further development of the F.3/48 should go ahead.

> The Air Staff felt that the need to develop this aircraft was important. The Hunter would be superior to the MiG-15, but not necessarily to a developed MiG-15, or a new Russian fighter. In addition, although the Hunter was

adequate against the current bomber threat it would need maximum possible performance against a jet bomber threat which might soon arise.[6]

The Air Ministry was right to be worried. The following year, the first flight of the Tupolev Tu-95 (NATO codename Bear) occurred. This swept-wing bomber was powered by turboprop engines, giving a useful performance of 520mph and a range of over 8,000 nautical miles at an altitude of 45,000 feet. In 1953 the Soviet Union flew its first pure jet bomber, the Myasishchev M-4 (NATO codename Bison), a swept-wing jet with a range of some 3,000 nautical miles and top speed of 588mph at an altitude of 36,000 feet.

With agreement that the Hunter should be developed without delay, on 26 February 1952 the Hawker RTO wrote to the company with arrangements for manufacture of the 50-degree development of F.3/48, leading on 18 April to the receipt of a draft specification for P.1083.

Notwithstanding the brochure suggestion of the development aircraft being fitted with an Armstrong Siddeley engine, on 15 May Messrs Herd and Kerry visited Kingston to discuss installation of the more powerful Rolls-Royce RA.14 with reheat capability into P.1083. While this was entirely feasible, Hawker Design was concerned that such an installation would require rather more fuel than presently available and, with the existing airframe, space for this

Hunter XF310 at Farnborough carrying 2 x Fireflash missiles.

Hunter XF378 at Dunsfold carrying 2 x Firestreak missiles.

addition could not easily be found. At a further meeting with Air Ministry and DOR staff in June 1952, held to discuss the development programme, the MoS representatives disagreed with Hawker's estimate of July 1953 for first flight of P.1083 and opined that, due to the conflict between the Hunter and the Sea Hawk production, the company would be in 'production difficulties for at least twelve months. They thought the developed Hunter could be delivered off production during the summer of 1955' – less than the good news that DOR was looking for.

Specification F.119D was issued in August 1952 to cover the prototype. In October, metal was cut for the first time for P.1083 and construction of the wings began in the experimental department at Kingston upon Thames. The aircraft was allocated serial number WN470 and ordered under contract 6/Aircraft/6296/CB.7b. A mock-up was also constructed and on 15 February 1953, a mock-up of the Rolls-Royce Avon RA.14R engine was installed but, as noted earlier, Hawker was now struggling to find space for the additional fuel required for the reheat version of the Avon engine.

At about this time, it was clear that all was not well with the basic Hunter and various shortcomings required rectification prior to entry into service. It became apparent to the Air Staff that the unsatisfactory state of affairs on the *existing* Hunter was an indication of the difficulties which confronted the Hunter development, quite apart from the aerodynamic changes involved in fitting a larger engine (also reheated) and the complication attendant on the fitting of guided missiles (Blue Jay). It therefore seemed clear that the Hunter development must be virtually a new aircraft and was unlikely to be available to the service within two years after the first Hunter, as originally planned. Because of the delays inherent in getting the developed machine into service and the additional costs that Hawker was now highlighting (£130,000), the conclusion within DOR was that, to allow Hawker to concentrate on clearing the Hunter F.1 for service, 'the Hunter Development should be cancelled in favour of the Swift Development which was scheduled for introduction into the Service in 1956'.[7]

This proposal was submitted and discussed at a meeting under the chairmanship of DCAS on 29 May 1953, a decision being made that 'the Hunter Development (Hawker P.1083) should be cancelled, and reliance placed on the development of the Swift (which must be tailored to take Blue Jay)'. Ironically, while it looked like the Swift had stolen a march over the Hunter, the Swift development was itself cancelled in late 1954. It was therefore in complete ignorance that Hawker Aircraft hosted a visit on 10 June 1953 by Air Commodore Silyn-Roberts (DOR) and Air Commodore Wallace Kyle (DMARD) at Dunsfold to discuss with Hawker the application of larger engines to the P.1067 Hunter. Quite what was said is unknown but, shortly after this meeting, on 22 June, Silyn-Roberts and Kyle were back, this time with Air Vice-Marshal Tuttle (Comptroller of Supplies, Air – CS/A) to break the news that P.1083 was no longer required, progress with larger engines being such that required thrust would be available without reheat and, on 13 July, the company was officially notified of the cancellation of the P.1083 requirement, Tuttle stating in August that the company should now proceed with a large engined non-reheat version of the Hunter.[8]

Such a design began under the project code P.1099, featuring the Hunter with the uprated RA.14 Avon engine, though a revised brochure was sent out in September 1953 to Woodward-Nutt, perhaps illustrating the P.1083 with the dry Avon engine, without reheat. With the P.1083 aircraft 80 per cent complete, all work stopped, but most of the fuselage sections were retained and formed the basis of what would become the Hunter F.6 with the 200 Series Avon engine rated at 10,500lb thrust, the prototype P.1099 being serialled XF833. Thus, the UK's chance of possessing a supersonic fighter in its inventory would have to wait till the 1960s and the advent of the English Electric Lightning.

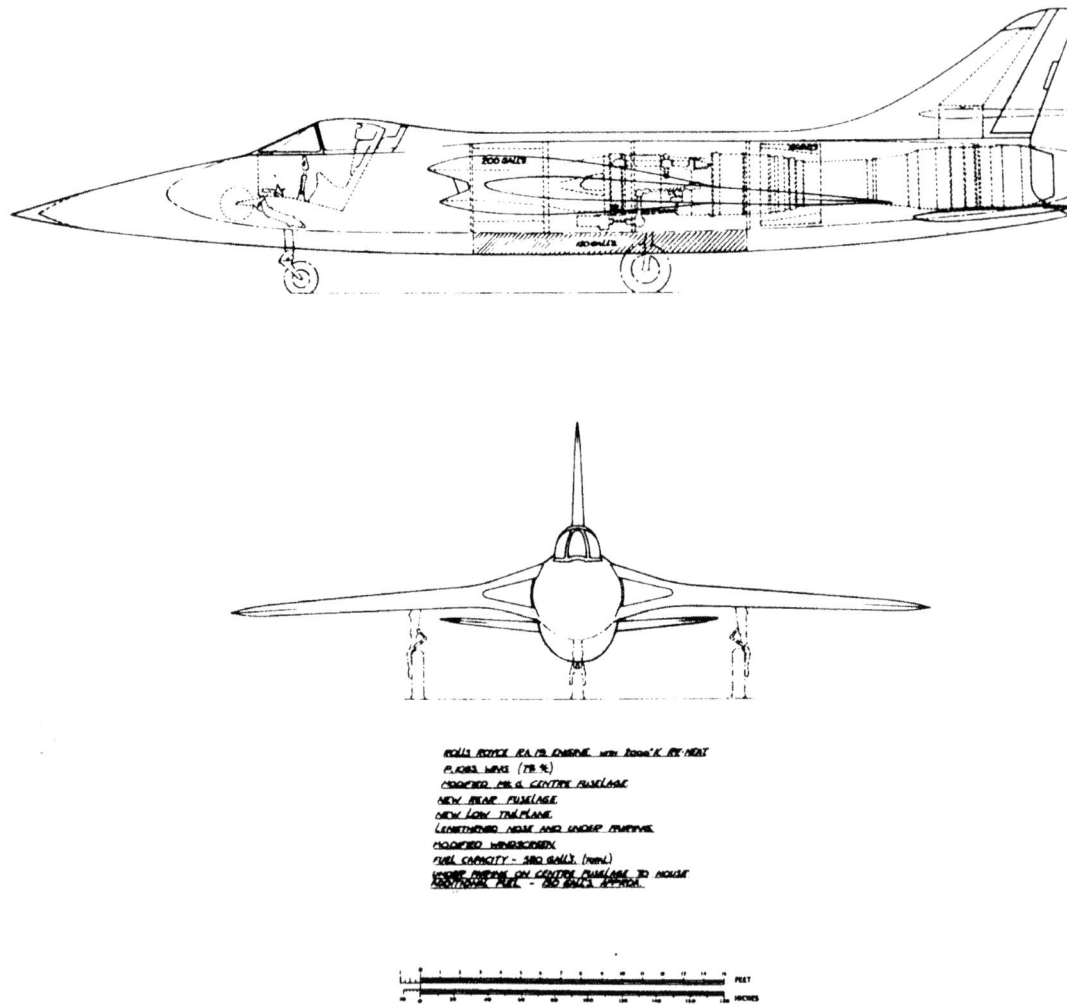

P.1083 GA Hunter development with 50° swept wing.

The P.1083 was the closest that Hawker Aircraft came to producing a supersonic Hunter development. However, the basic Hunter airframe would continue to be the subject of more minor developments to produce greater engine power, changes to the wing to increase aerodynamic efficiency and improvements to weapon carriage to allow the aircraft to function as a ground-attack platform. Notwithstanding this work, the basic design parameters were constantly in front of the Project Office team as they sought to achieve something more potent from the aircraft.

The concern regarding shortage of fuel in the P.1083 design was revisited in a design dated spring 1955 and coded P.1083-8, which featured a radical redesign

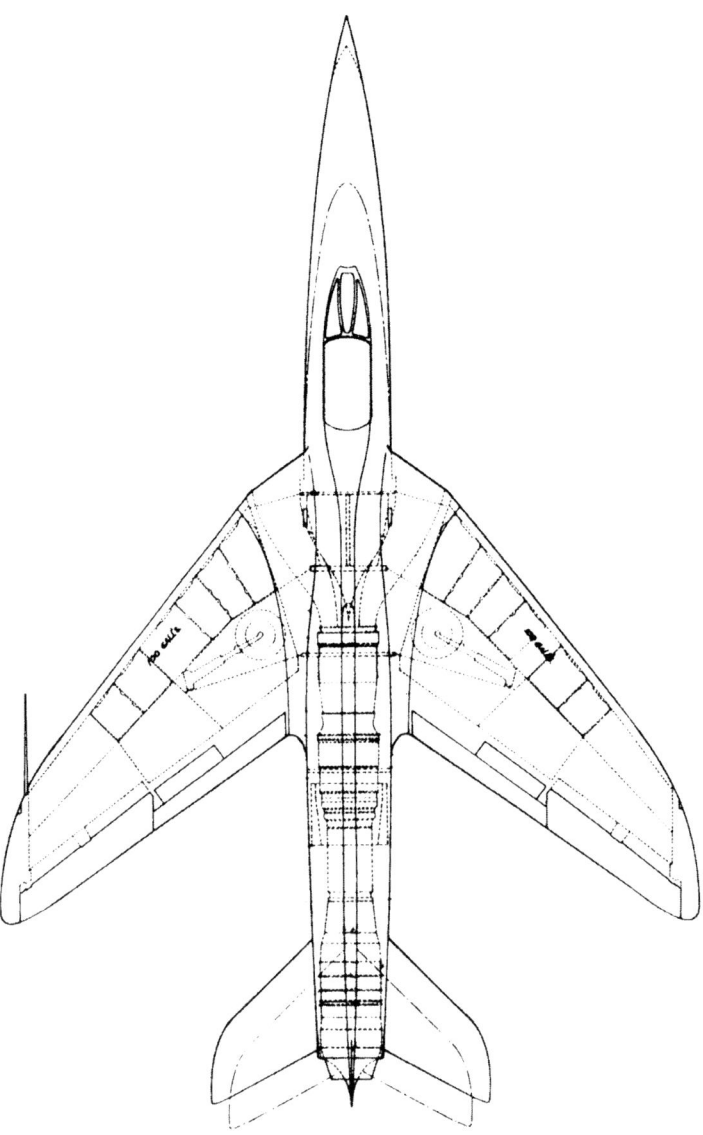

of the basic Hunter shape, including a modified centre fuselage providing a deep keel section to accommodate an extra 150 gallons of fuel. Other changes were made to the engine (Rolls-Royce RA.19 with reheat), a modified longer nose section, presumably capable of housing an AI radar unit and modified rear fuselage incorporating a lower tailplane position. Nothing came of this design, nor did the suggestion to install a reheated de Havilland Gyron engine. This design, the P.1090, required rather larger intakes which were carried forward of the wing leading edge to achieve this though the proposal did not proceed.

The Delta-Winged Hunter

The means by which Hawker sought to extend and develop the basic Hunter performance then were several. The use of increased wing sweepback and larger engines has been discussed above, but also considered were differing wing planforms, one of which, the delta, was becoming more widely accepted within the aviation community.

The delta wing had a number of attractive attributes which led to it being used by two of Hawker Aircraft's stablemates in the Hawker Siddeley Group, Avro with the 698 Vulcan medium bomber and Gloster with the Javelin. While increased wing sweep was valuable in allowing an increase in speed, because of the risks of flutter and aero-elasticity, the resulting structure tended to be rather heavier than the equivalent straight wing. The delta wing avoided this problem by having a much longer root chord (and therefore a lower t/c ratio), the greater wing area giving excellent rigidity, allowing for shorter span and thus allowing a lighter structure. This also allowed for an increase in the fuel volume that could be accommodated in the wing. A substantial leading edge sweep coupled with the delta planform permitted a low drag design capable of high subsonic or supersonic speed and drooping the leading edge could increase lift without a consequent increase in drag.

Work began in April 1951 on preliminary investigation of delta wing design, project numbers P.1084 and P.1085 being allocated. This was followed in October by a Fozard scheme coded P.1091 in association with Avro, a Hunter with a 60-degree delta wing powered by a Rolls-Royce RA.14R or 8,000lb AS Sapphire 4 reheated engine. Fitted with the later 200 Series Avon engine, the design should have been capable of level supersonic flight, though it was another design that failed to win friends. The following year, P.1092 and P.1093 were delta fighter designs for two-seat and single-seat aircraft respectively. In May 1953, in response to requirement ER.134T calling for a highly supersonic research aircraft to investigate this region of flight, P.1096 was another attempt at a delta (more correctly semi-delta) layout, with low-set tail and powered by a Rolls-Royce RB.106 engine and P.1097 with swept wing and T-tail, but, again, no further work was carried out. While occasional delta-winged designs would surface in the fertile minds of the Project Office, Hawker would never produce a delta-winged aircraft (unless one counts the wing of the early P.1127 VTOL aircraft).

In April 1953 the idea of a supersonic Hunter was back on John Fozard's drawing board at Kingston, designated P.1100. While retaining the 40-degree wing sweep and tail unit of the basic Hunter planform, a new pointed nose would accommodate an AI.20 radar while 130-gallon wing-tip-mounted tanks would give a useful increase in fuel load. While the engine was not specified, two rocket motors of the AS Snarler type were included in the trailing-edge wing roots. Armament would comprise two 30mm cannon in the nose and Blue Jay missiles under the wings, these being the DH Firestreak infra-red homing missile which was the UK's first guided missile in service with the RAF and Fleet Air Arm.[9]

Fozard tried again in October 1953 to get a performance increase out of the Hunter airframe, this time by using a thinner wing coupled with a Rolls-Royce RA.19 engine with reheat, the scheme being designated P.1102. The main undercarriage was moved out of the wing and into the fuselage. While the t/c ratio is not mentioned, one must assume that it was thinner than the Hunter due to the need to relocate the undercarriage and the increased wing sweep of 45 degrees would have had a beneficial effect upon t/c ratio. Two 30mm cannon were retained in the nose and side-mounted airbrakes installed on the aft fuselage.

Yet another attempt to coax extra performance from the Hunter was P.1105 of March 1954. While retaining the standard Mk.6 layout, two Napier TRR/37 rocket packs were carried under the wings on pylons. Fuel for the Rolls-Royce Avon RA.23 amounted to 330 gallons and for the rockets 60 gallons, plus 310 gallons of oxidant.

This was later followed by P.1106, another 'thin-wing Hunter' scheme but this time using the AS Sapphire Sa.10 engine. Wing sweep was retained at 40 degrees but the t/c ratio slimmed down from 8.5 per cent of the standard wing to 6 per cent, while span was increased from 33 feet eight inches to 38 feet. The aircraft was drawn with two Blue Jay (Firestreak) missiles carried under the wings but no cannon in the nose. This scheme was refined the following year by J.D. Mills, now showing a nose-mounted AI.20 radar system for target acquisition, the engine now a Rolls-Royce Avon RA.28.

The All-Weather Hunter

Later development of fighter aircraft would see the single-seat missile-equipped fighter become central to the inventory of most air forces, but, for Hawker, the time was not sufficiently ripe for such a proposal, Hawker's all-weather schemes revolving around two-seat fighters.

66 Hawker's Secret Projects

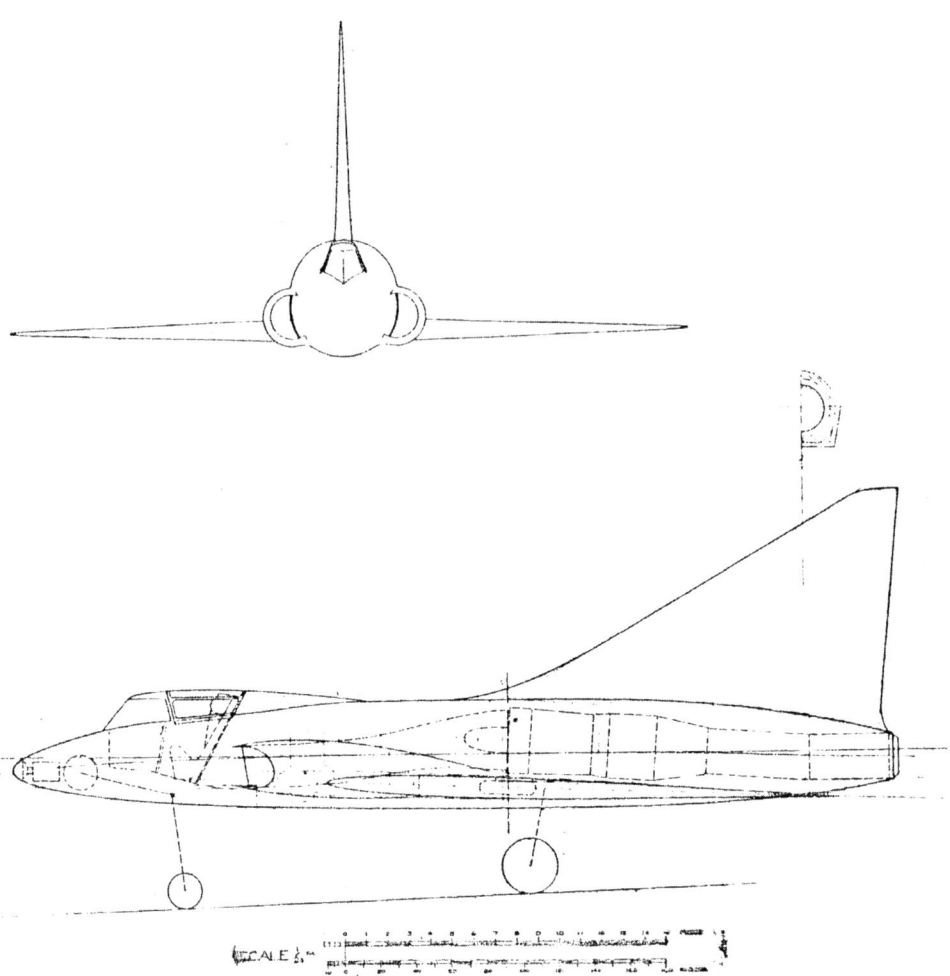

P.1084 GA delta Hunter development.

As the understanding of the complex utilisation of the air intercept radar/missile package matured, it became increasingly clear that pilot workload would be onerous during the attack phase and that therefore a two-man crew with one dedicated to the weapon system might be preferable. To this end, various schemes were drawn using the P.1101 Hunter T.7 two-seat trainer as the basis for an all-weather fighter. The first stab at this layout, in November 1955, was P.1114 powered by a Rolls-Royce Avon 203 and mounting an AI.20 type radar in the nose. Wing-tip-mounted fuel tanks gave a useful additional 75 gallons a side. No missile armament was specified and cannon complement was reduced to just two 30mm Aden units. This initial scheme was refined by A.E. Barrett

Hunter Developments 67

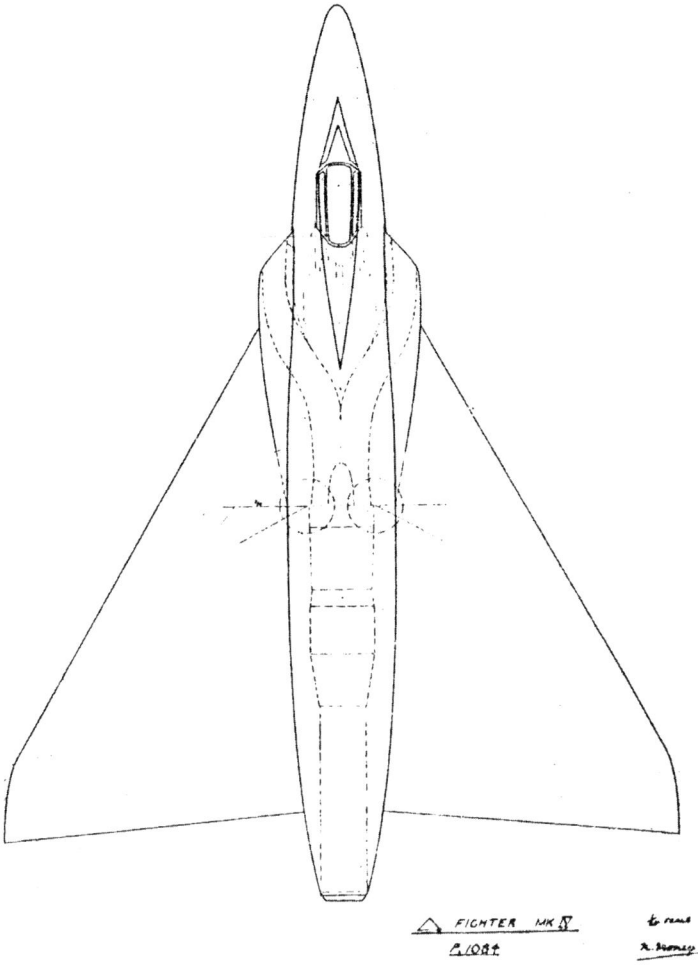

in January 1956 to include underwing-mounted Firestreak-type missiles and side-mounted airbrakes, both Avon and Sapphire engine types being suggested. There was even a stab at a naval version by Ralph Hooper, the P.1117 reverting to single-seat layout, retaining the wing-tip tanks and intercept radar/Firestreak combination and adding an arrester hook. Whether carrier operation was ever investigated in detail or whether the hook would have been for land use only, is unknown but would certainly have been a new hunting ground for the Hunter.

Interestingly, even when Hawker Aircraft Ltd had moved on to studies for a Mach 2 interceptor, investigation of all-weather Hunter variants continued. Whether this was a reaction to problems in 'selling' the P.1121 to the Air Ministry (investigated in the next chapter) is unknown but it may well represent a 'casting about' for projects to fill the void in the Project Office generated by the company decision to eventually terminate the project. Alternatively, it may represent a response to the continuing demands from government that new aircraft should be integrated into the weapon-system concept, as seen with the P.1121 project, at a time when the Sandys White Paper had deliberately pointed the way to the demise of the 'dumb' gun-armed interceptor. Certainly, the Hunter would have provided a cheaper alternative to a completely new design, though, in performance terms, it was by now already looking long in the tooth.

P.1130 drawn by A.E. Barrett in September 1957 started again with the basic T.7 trainer layout, now fitted with a more powerful Avon 203 engine and AI.23 type radar and with retention of two Aden cannon plus provision for two Blue Jays. This was worked up following interest in the concept from the Indian Air Force. It was completed in December 1957 and taken by the hand of Frank Murphy in January 1958 to India for discussion. However, it appears that the cost and complexity of the aircraft package shook them somewhat and they requested that some of their single-seat Hunters be modified to carry Firestreak, though this went no further. Then, in July 1959, the company was contacted by the Indian Air Advisor, Group Captain Moolgavkar, regarding renewed interest in the concept and detailed proposals were furnished to the Indians but, again, nothing further transpired.

Finally, with a reversion to single-seat layout, Hawker's last throw of the all-weather Hunter was the P.1133. This came with a AI.23 radar and twin Firestreak or Fireflash missile carriage or, alternatively, omission of missiles and replacement with the full complement of four Aden cannon, and the P.1135, by J.D. Mills in January 1959, another 'thin-wing' design (6 per cent t/c ratio), now powered by a reheated Avon RB.146 engine giving 13,200lb thrust dry and 18,300lb 'wet'.

Thus ended Hawker's attempts to develop the Hunter into a more potent interceptor. The investigations had continued for over a decade which saw the basic Hunter becoming more and more outdated as an interceptor. Fortunately for the company, early work to suit the aircraft as a ground-attack platform was rewarded by the Hunter FGA.9 becoming a formidable platform for this task and enjoying export around the world. But before we leave Hunter development behind to pursue the P.1121, one last throw of the dice produced the most unusual Hunter of all.

The Executive Hunter

The P.1128 originated on the board of John Fozard in 1957 and represented another of those attempts to fill the P.1121 void. This was Fozard's thoughts on a Hunter six-seat executive jet. Powered by two Bristol Orpheus engines, the design borrowed heavily from the Hunter, the wings and tail being wedded to a stubby cabin into which were the passengers were shoe-horned. The engine intakes appear to have been mounted either side of the upper fuselage behind the wing, their former positions in the wing roots being replaced with fuel tanks. It is difficult to know whether this alteration was pure research into alternative

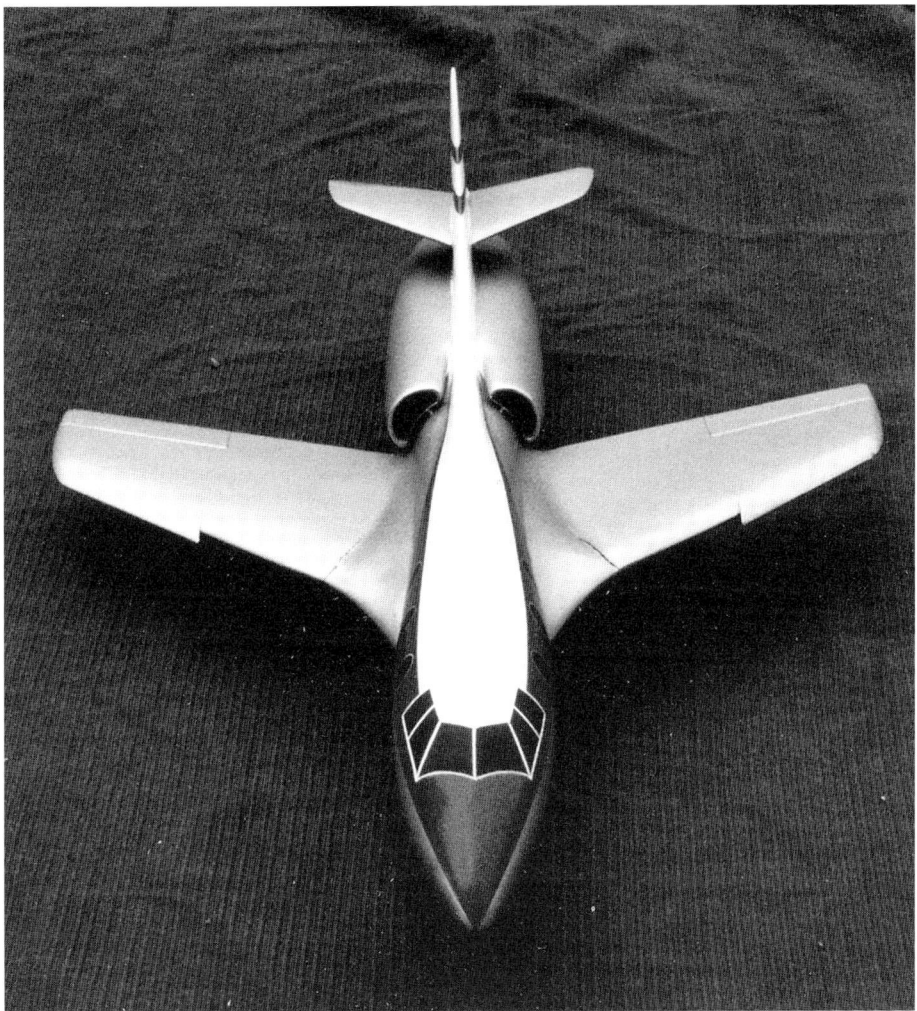

P.1128 Hunter jet transport development model.

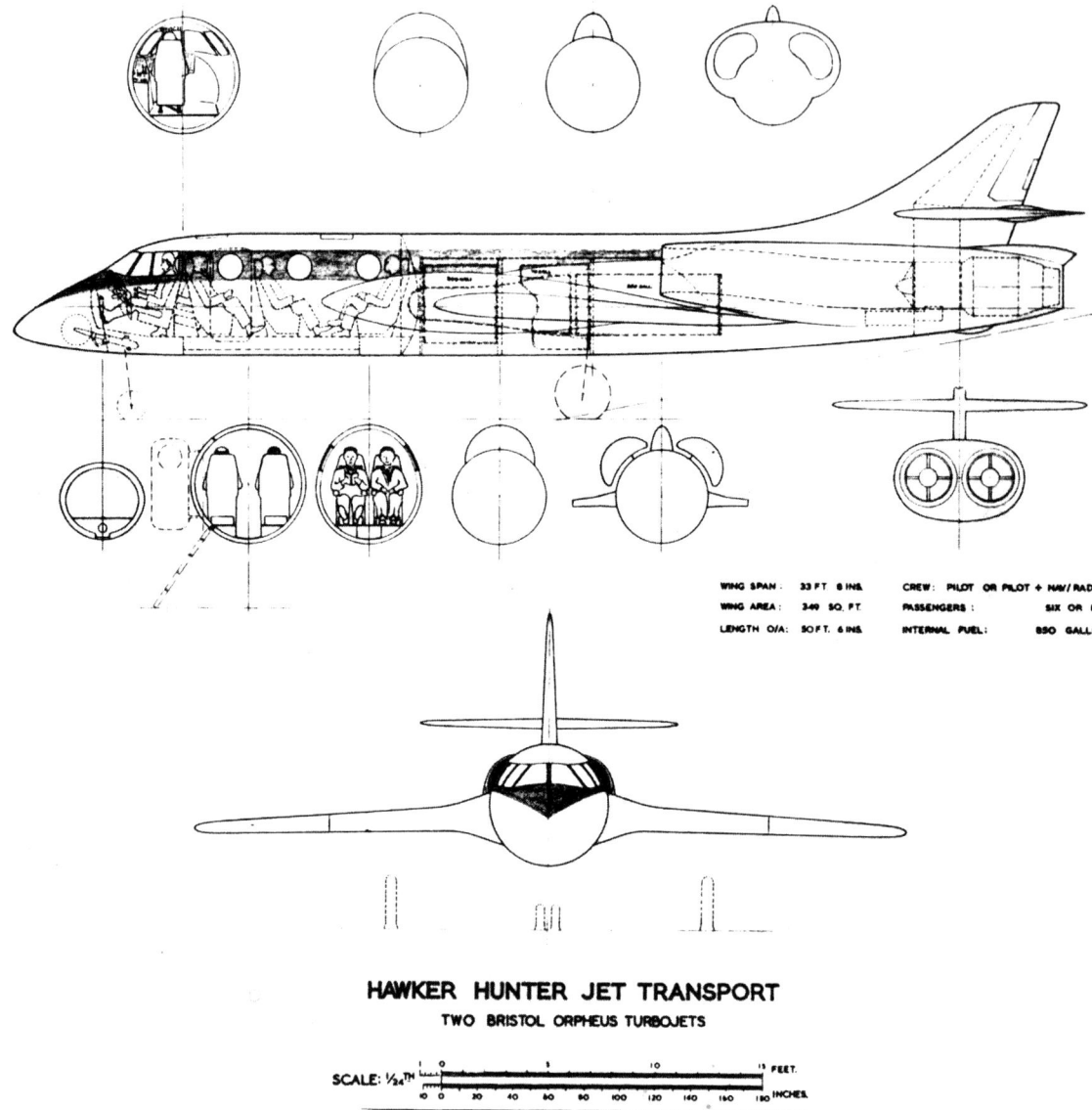

P.1128 GA Hunter jet transport development.

uses or mere desperation, but Mr Fozard would produce throughout his time in the Project Office some unusual layouts, none of which progressed very far.

With the benefit of hindsight, it can be seen that the decision by the Air Ministry to cancel the P.1083 was probably the right one; the problems with providing sufficient fuel to give a useful flight duration were not being solved with the existing layout. That said, the Avon 200 Series engine at 10,000lb

Hunter Developments 71

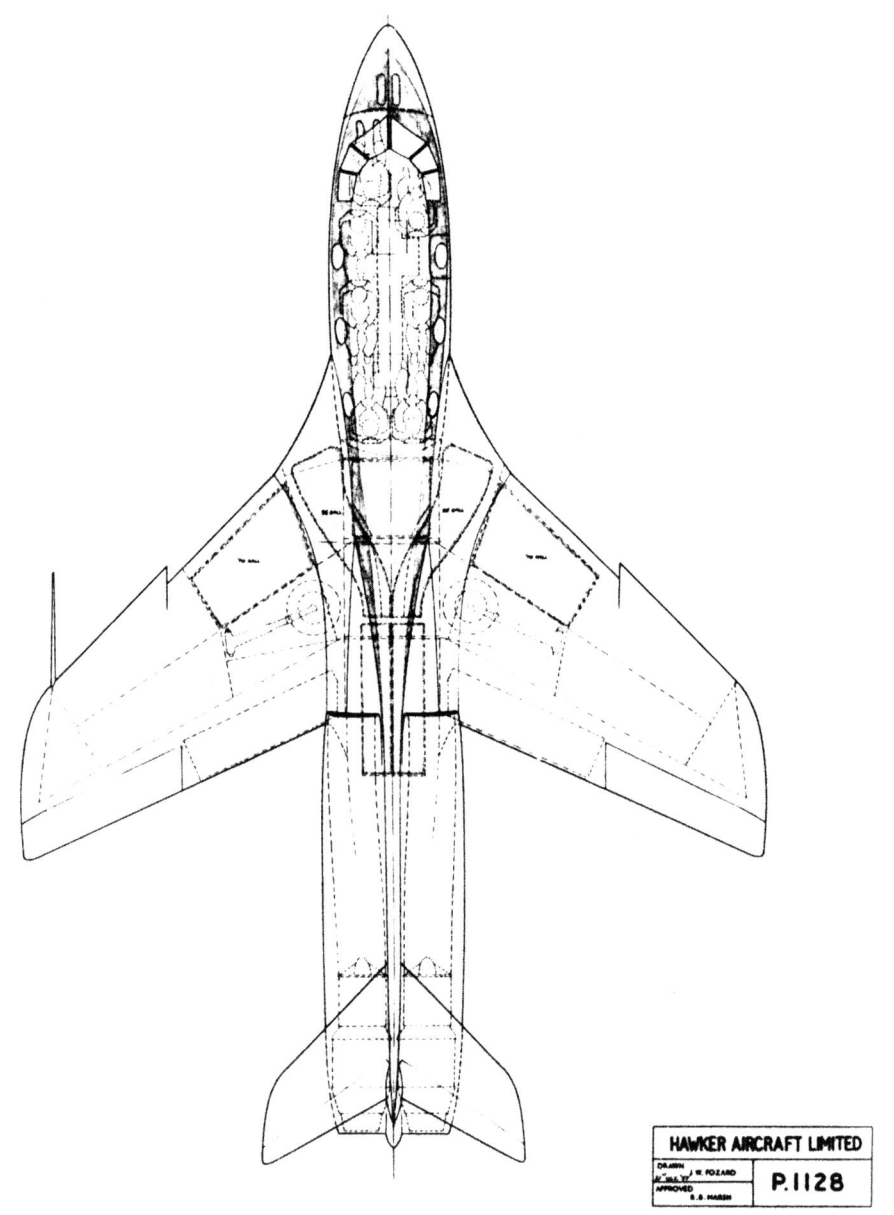

thrust coupled to a 50-degree swept wing would have provided a performance comparable to the Mk.3 Hunter with RA.7 Avon and reheat, but capable of modest supersonic speed in level flight. When the Avon 200 was fitted to the Hunter Mk.6, the aircraft was still only capable of supersonic speed in a dive, the wing being the limiting factor, though it did allow the later FGA.9 ground-attack variant to carry a very useful tactical weapons load with little performance loss.

To a great extent, the latter days of attempts to wring new life out of the Hunter design were overshadowed by Hawker Aircraft's 'next big thing'. This was a large supersonic fighter that would pass through many iterations before eventually foundering on the rocky shores of the 1957 Defence White Paper promulgated by the then Minister of Defence, Duncan Sandys. This project is examined in the next chapter.

Chapter 4

The Search for a Supersonic Fighter

The quest for speed which obsessed and fascinated western society in the post-war period reflected the pre-war fascination with high-speed transport on land and in the air and its proponents were lauded as heroes. Speed had become almost an end in itself. In aerial warfare, while a speed advantage over adversaries was almost a prerequisite for success, it became almost the *sine-qua-non* of post-war military aviation thinking. In the UK supersonic flight was investigated during the dark days of the Second World War, Miles Aircraft being allocated a research contract for a supersonic aircraft, the Miles M.52. This attempt to produce a vehicle with which to break the 'sound barrier' unfortunately fell on the altar of politics, Ben Lockspeiser, Chief Scientist at the Ministry of Supply, announced the project's cancellation in February 1946, ostensibly on grounds of pilot safety, though actually on those of cost. The US was, predictably, the first nation to break the 'sound barrier' when Chuck Yeager flew the rocket-powered Bell X-1 to supersonic speed on 14 October 1947 in an aircraft not dissimilar to the M.52.

Henceforth, air speed above that of sound (i.e., Mach unity) became the goal of the leading nations, with less regard for how this might actually be used in practice or concern regarding the compromises required in other areas, and more regard for a perceived 'need' to be faster than the opposition. Within the Air Staff and the RAF, what reflective thought regarding supersonic flight occurred went something like this. 'We know that flight at supersonic speeds is possible and we therefore assume that the "other side" will be pursuing the technology required to produce supersonic bombers with which to threaten "our side", ergo, we must have supersonic fighters to shoot them down.' As far as it went, this was a perfectly proper stance to take, though the supersonic bomber would be many years in coming to fruition. Meanwhile, the western nations comprising NATO entered into an urgent investigation to design and produce a suitable counter to the perceived threat from the east just when the UK's financial situation was most fragile.

While supersonic flight was certainly possible – Hawker had flown the Hunter supersonically in June 1952 – the use of reheat to achieve it entailed a profligate use of fuel, reducing the already short duration of fighters at that time. Moreover, once interception had begun, successful attack could not be completed at supersonic speed, given the need to manoeuvre and bring one's weapons to bear on the opposition, this usually being the 20mm or 30mm cannon, aimed by the eye of the pilot. So, while supersonic capability was of great use in getting from A to B as quickly as possible so that acquisition could occur as far from UK shores as possible, to achieve this, vast sums of money would be poured into the aircraft industry that might more profitably have been used to improve acquisition and loiter times, and successful destruction of the target.

So it was that, with 'supersonic' the watchword of the day, Hawker had been pursuing their quest for a supersonic Hunter derivative – unsuccessfully. Projects were examined in the early 1950s, including one in response to operational requirement ER.134T, issued in December 1952, which called for an aircraft capable of Mach 2 plus to further research into high-speed flight and, in particular, the effects of kinetic heating at these speeds, to produce data applicable to Avro's Mach 3 bomber proposal. The Hawker Project Office (John Fozard) produced designs under codes P.1096, which showed a single-seat aircraft with high swept wing (56.5 degrees), all swept low-position tail surfaces and powered by a Rolls-Royce RB.106 engine (projected to produce 21,750lb thrust with reheat), and P.1097, a single-seat, swept mid-wing (52 degrees) with T-tail and, rather unnecessarily, radar and four-cannon armament! The contract was awarded to Bristol who produced their 188 stainless steel aircraft, first flown many years later after the cancellation of the Avro 730 supersonic reconnaissance bomber had rendered its primary task obsolete.

However, the opportunity had been provided to the Hawker Project Office to think about just what a Mach 2 fighter might look like. From these early thoughts emerged designs in early 1952 by Fozard for a 'single-seat supersonic fighter' featuring a blended delta wing and, in late 1953, for a supersonic research aircraft resembling the earlier P.1096. These initial design schemes were the manifestation of continued concern in the Air Ministry about just what the next fighter might comprise. Speed was considered of primary importance to enable approaching bombers to be intercepted as far from the UK coast as possible. Various consultations with the aircraft manufacturers would eventually lead to Air Ministry thoughts coalescing around a new operational requirement and its attendant specification being disseminated to interested companies.[1]

Receipt of OR.329 on 17 March 1954 allowed the Kingston design team to begin a more focused consideration of just how to respond to the requirement, a new project, P.1103, being instigated. This project would feature the new de Havilland Gyron engine then under development which offered Mach 2 performance, a twin-engined design coded P.1104 by Fozard being produced at the same time using Rolls-Royce RB.112 engines.

Initial concerns revolved around the likely layout; in July, three possibilities were being considered:

a) a low-wing with wing root air intakes
b) mid-wing arrangement with wing root air intakes
c) medium-low wing with a single under-fuselage air intake.[2]

The earliest drawings, of March 1954 (Option a) featured a two-seat aircraft of 62 feet and a Hunter style empennage, a 50-degree swept 37 feet span high-mounted wing of 4 per cent t/c ratio, powered by DH Gyron engine with reheat, fed by low-mounted D-shaped intakes in the lower fuselage. Armament was projected as twin cannon mounted above the intakes and carriage of unspecified air-to-air guided missiles and unspecified nose-mounted radar. Fuel of 960 gallons was to be provided, presumably in the fuselage since the thin wing would offer limited space for wing tanks. A low wing and leading-edge intakes (Option b) was detailed in a further drawing by Hooper, this version being termed P.1103 Mk.2.

In July 1954 the intake position was again examined and the wing-root intakes deleted. For the first time, a ventral intake of 'letter-box' shape was featured, possibly influenced by Robert Lickley (certainly a sketch drawing exists with his name attached). Around this new configuration the design had now grown to 67 feet with a span of 36 feet. Twin seats and Hunter-style empennage with wing trailing edge rocket motors made for a large and impressive aircraft.[3]

In October it was agreed that the single under-fuselage intake was the best arrangement. Meanwhile, meetings were organised with various engine and radar experts to begin the process of agreeing the avionics and engine choices.

While work continued on the large two-crew version, in December 1954, possibly in response to advice from RAE which preferred a smaller single-seat interceptor, Hooper schemed a single-seat aircraft with small radar and ventral intake replaced by wing-root intakes feeding a 10,000lb thrust engine. Length would be only 47 feet and span a mere 28 feet. He also drew a slightly larger but similar layout around the same time with length of 57 feet and span of 32.5 feet; the engine would be in the 14,000lb class; both these designs featured wing trailing edge root rocket motors.

The receipt of specification F.155T on 15 January 1955 confirmed the new requirement's salient points: the ability to intercept high-altitude high-speed bombers approaching the UK in all weathers. A two-man crew of pilot and radar operator/navigator would have the capability of achieving 60,000 feet and Mach 2 flight within seven minutes of brake release. Armament would be air-to-air missiles; collision course interception would be covered by development of the Vickers Red Hebe radar-guided missile and pursuit course attack by development of the DH Blue Jay IV (later named Blue Vesta and entered service as Red Top).[4]

The specification aroused much interest within the UK aviation manufacturing concerns, including Armstrong Whitworth and Avro (both Hawker Siddeley Group members), English Electric, Vickers, Fairey, Saro and de Havilland all submitting proposals. With receipt of the OR and specification, the Hawker Project Office work already completed was swiftly aligned with the new requirements, the emerging design being drawn around later iterations of P.1103.

With the receipt of Issue 2 of F.155T, the Project Office sought to fully answer the requirement with a new design reverting to twin-seats, a large search radar and twin guided missiles. The length had settled at 63 feet, span 39 feet and the ventral intake had been enlarged to allow use of a 20,000lb thrust engine for which suggestions included the DH Gyron, Rolls-Royce RB.122, Armstrong Siddeley P.173 or the Canadian Orenda PS.13. Internal fuel of 1,100 gallons could be supplemented by external tanks offering a further 170 gallons per side. Thrust was to be further boosted by self-contained rocket motors mounted in streamlined pods on each wing at mid-span, powered by a kerosene/hydrogen peroxide fuel. The wing was drawn as a four-spar arrangement, the spars cranked at the root, with the trailing edge accommodating split ailerons and plain flaps. Fuel tankage would be integral rather than the use of bag tanks. The tailplane was drawn both on the lower fin position as per the Hunter and as a low fuselage location below the jet pipe, though the latter would require a taller undercarriage to ensure clearance for the tailplane on take-off. A further drawing featured the Ministry's preferred radar-guided missile, the Red Hebe mounted at the wing tips, though the size of the missile (up to 22 feet long and 1,330lb/600kg) in this location was far from ideal. Work was also carried out to design retractable intake guards due to concerns about the ventral chin intake being susceptible to FOD.[5]

The tenders to the specification for this ambitious fighter were expected to demonstrate that the companies had absorbed the need to view the resulting aircraft as part of a complete integrated weapon system which included all the equipment required to carry out its intended role, i.e., navigation, radar, fire control etc. To this end, meetings were arranged with government civil servants and scientists at the research centres to advise and guide the various design teams in the choice and availability of equipment.

At a meeting with Air Ministry representatives from the research and development department, it was accepted that carriage of four weapons (i.e., two radar-guided, two infra-red guided) plus the control equipment for either active or passive weapons in the same aircraft would be dropped. Interestingly, at this same meeting, Hawker gained the impression that the civil servants did not believe that any company would be able to fulfil the specification and it would therefore be a case of examining the various compromises put forward. At another meeting, in March 1955, this time with officials at the Royal Radar Establishment (RRE) at Malvern, to try to obtain information regarding the available radars for F.155T, it was stated that, in the opinion of the AI (Airborne Interception) section at RRE, the requirements of the specification could not, and should not, be met. No doubt having shocked his audience, Mr Atherton continued by explaining that putting the aircraft derived from the specification into service too early would mean that its radar would be much heavier, lack range, or both. Much better to aim for an in-service date of 1965 and use the intervening period to produce an outstanding lightweight, powerful radar with which to answer the specification's requirements, having a range of some 40 miles for a smaller dish and a weight of just 500-600lb. However, at a further meeting in June 1955, Mr Atherton explained that the situation was now somewhat clearer, RRE focusing on an AI providing a 20-mile range and aiming for in-service in small numbers in 1962.[6]

Most of the initial P.1103 design work was Ralph Hooper's. Meanwhile, John Fozard had been quietly working away at another design that broadly followed the F.155T specification but with a significantly different layout, coded P.1104. Initially schemed in January 1954 with two RB.112 engines, this matured into a twin DH Gyron Junior design with 2,000°k reheat, the engines mounted in underslung pods on the wing. The wing was swept at a modest 37 degrees and a t/c ratio of 4.25 but the tailplane was not swept. The 'mark 4' version now

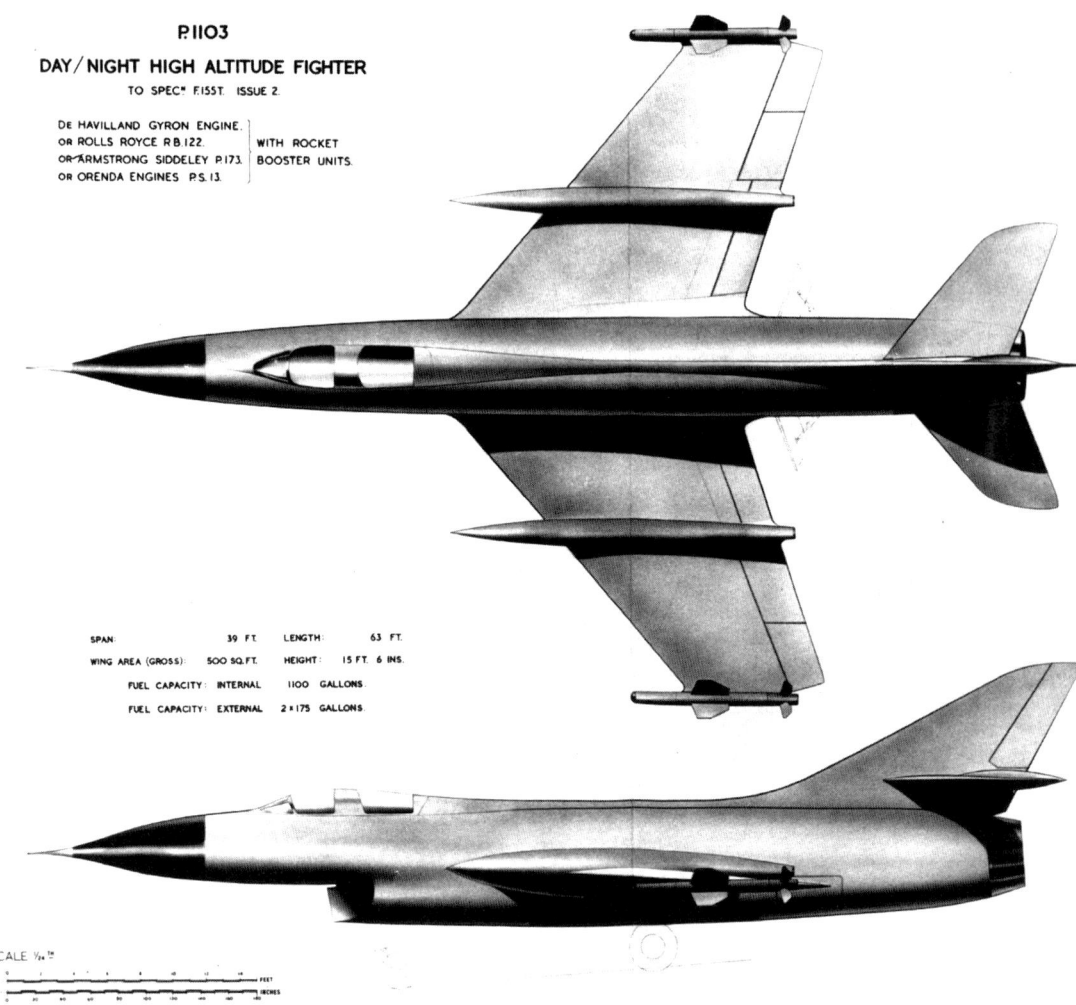

P.1103 GA, all-weather, high-altitude supersonic fighter.

sported twin cannon in the lower nose and missiles carried in an under-fuselage position below the cockpit while the tailplane had acquired noticeable dihedral and moderate sweep.

Fozard followed this up in July 1954 with an adventurous design titled P.1107 which featured a length of 69.5 feet, a thin tapered wing of 37 feet and no less than six RB.115 engines in underwing pods carried at the wing root, plus an additional rocket motor of 4,000lb thrust in the tail. The shoulder-mounted wing resulted in a bicycle undercarriage with outriggers at the wing tips. Further refinement of the layout resulted in length reducing to 63.4 feet, wingspan to 36 feet and a modest sweep of 26.5 degrees. Engine numbers were reduced to

The Search for a Supersonic Fighter 79

four and the outriggers brought inboard to the engine-pod location; another drawing showed the four engines replaced with just two mounted under the wing and integral with the fuselage. By the spring of 1955, Fozard's design for P.1107 had returned to underwing engines mounted at mid-span, bolstered by a tail-mounted rocket motor but appears to have proceeded no further, all effort now, no doubt, being directed to refining Ralph Hooper's P.1103 final design for submission.

Throughout the various different schemes examined, one emerging constant was the plan to use the DH Gyron turbojet engine. In the early days of the P.1103 proposal this was really the only engine available which promised the high power required for Hawker's single-engined approach. Though other engines were considered, most were still just 'paper' engines with all of the uncertainty that came with them. The Gyron had its genesis in a specification, TE.6/50, of March 1950 for an engine suitable for a supersonic fighter. De Havilland decided to take a leap from their current modestly powered Ghost and Goblin engines to a unit capable of producing 20,000lb thrust though at that stage (1951) there was no government funding available; this was addressed in 1953 and funding was then forthcoming. Ultimately, its development having followed the development of the P.1103 and later the P.1121, both the engine and the airframe would fail to enter service.

Finally, on 5 October 1955, Hawker submitted the tender for the F.155T specification. This featured a Mach 2 performance from a normal light alloy construction, a developed AI.18 radar system and DH Gyron engine. It is worth noting that a lot of work and research went into this design at Kingston, and that 'advice' from government civil servants and Air Ministry sources had at times been less than clear and unambiguous, the result of which would tell in the subsequent examination and debate within the specialist Air Ministry departments given the task of assessing the various submissions from industry.

Assessment of the responses to F.155T had seen concerns that the Hawker submission would not meet the specification, though further amplification of the design was supplied late in the year, documents being submitted clarifying the Hawker response to the requirement to carry Red Hebe. Also, in January 1956, further drawings were submitted demonstrating carriage of (US) Sparrow and Falcon missiles as, presumably, lighter weight alternatives to the UK missiles. However, it was to no avail. In January 1956 the F.155T committee decided to eliminate Hawker and English Electric, though this was not communicated to the company, which continued work on refinement of the design, particularly wind tunnel analysis at Armstrong Whitworth. In March the committee recommended that the Fairey and Armstrong Whitworth designs be developed further and, in April, Hawker was belatedly informed of their failure to win the project.[7]

Comments from government and specialist assessors ranged from the aircraft being too small, tandem rather than side-by-side seating, poor avionics access, conservative performance and with an engine of limited development potential. It is probably fair to say that, whereas Hawker had worked to design an aircraft that could actually be built with current expertise and enter service with the minimum of development, Fairey, with a steel and titanium design, and Armstrong Whitworth, with an extremely thin fuselage akin to the Bristol 188, had provided schemes which would in reality have required many years of development before there was any chance of their entering service. In the event, no service aircraft would ever emerge from F.155T, which was perhaps just too ambitious for a direct leap to its requirements from the then current state of play.[8]

On 3 May 1956, shortly after learning that the P.1103 project was a non-starter (and it is fair to presume that Hawker expected to win the tender), Camm paid a visit to DCAS Sir Thomas Pike to discuss the future and learned of a probable Air Staff requirement for a long-range interceptor with ground-attack capability. As the Hawker day diary put it 'Since the OR.329-type fighter was unlikely to fulfil this general-purpose role, DCAS suggested that Hawkers could profitably adapt the P.1103 to this role'. This was followed by meetings with de Havilland Engines to discuss improving the specific fuel consumption of the Gyron engine to suit the strike role of P.1103, and with Handel Davies, Chief Scientific Officer to the Air Ministry, to discuss probable requirements for a P.1103 strike version and to obtain details of the nuclear weapon which it would be required to carry. On 8 May DGTD, E.T. Jones visited Camm to discuss the future, Hawker believing that 'he gave encouragement to the P.1103 strike version'.[9]

With this apparent encouragement from government sources, at Kingston, Camm's Hawker team were sufficiently knowledgeable to see that the 'winning' designs to F.155T would not result in a fighter entering service in the timescale specified and confident that their design was more than capable of being eventually accepted as the fighter that the RAF really needed, whether the 'experts' at the OR branch of the Air Ministry could see it or not. Following the discussions noted above and with Hawker Siddeley board executives, Camm felt that he had sufficient backing to continue pursuit of a contract for a fighter/strike aircraft based on the P.1103 and instructed the Project Office accordingly. However, from now on, all work would be on a private venture basis – the company would be paying the bills.

P.1116

Accordingly, on 8 May 1956, project P.1116 was instigated for a 'Mach 2 Interceptor and Long Range Strike Fighter' powered again by the DH Gyron engine. Fozard now took up the cudgels rather than Hooper, his design influenced heavily by the P.1103 but depicting a single-seat aircraft with twin 30mm cannon installed adjacent to the cockpit. Length was 63.5 feet and span now 32 feet. Internal fuel capacity was increased to 1,360 gallons, with a further 250 gallons per side an option in wing-tip tanks. Bombs or guided weapons were seen as the most likely stores and would be carried under the wing. All in all, another graceful Hawker design from this experienced design team. Just eight days later, Camm was back with DCAS Sir Thomas Pike with new brochures detailing the P.1116. However, second thoughts on the new design quickly saw a retraction of P.1116 since it scrapped some design work and construction already completed for P.1103. Therefore, a new design study using more of the work already put into P.1103 resulted on 28 May in a new designation requiring a minimum of structural changes; this was designated P.1121, upon which most subsequent work would be carried out.[10]

P.1121

P.1121 was described as an Air Superiority Strike Fighter (ASSF), a single-seat design with a length of 65 feet 9 inches and span of 37 feet. Fuel capacity was to be 1,500 gallons internal and the capability of another 600 gallons carried in four external tanks. Armament was envisaged as fifty 2-inch unguided rockets carried in pop-out packs behind the cockpit augmented with further rockets, missiles or bombs carried under the wing. Power was again to be provided by the DH Gyron engine offering in the region of 20,000lb-plus thrust and landing speeds were addressed with plain flaps, with airbrakes either side on the rear fuselage and another in the ventral position. A search radar was allocated to the nose and the dihedral tailplane was to be located in the low fuselage position.

P.1121 in typical RAF paint scheme.

Over the following six months the many visits by various RAF and Air Ministry personnel offered conflicting advice on whether interception or strike should be the foremost role for the design, though the point was made that funding for fighters was likely to be scarce for the foreseeable future. A prime concern was to answer the requirement to provide reliable navigation for the low-level role and an inertial navigator was considered the best response to this. Unfortunately, Sperry, in whose hands the development work in the UK lay, had virtually stopped work on the instruction of the Air Staff, with the result that a second crew member was now considered to deal with the navigational aspects of the role by more traditional means.

A local board meeting on 31 August 1956, confirmed that work would proceed on P.1121 as originally conceived, i.e., single-seat strike fighter with low-altitude ground attack as a strong secondary role. At this point discussion began on provision of a prototype aircraft with the minimum of installed equipment with a target date of April 1958 for first flight. De Havilland agreed to concentrate on producing an engine for the prototype rated at 18,000-23,000lb thrust with reheat, with the earliest testing date being January 1957. In October 1956, during a visit by Air Marshal Satterley ACAS/OR, Hawker learned of initial work being carried out on a new operational requirement for a Canberra replacement and the possibility of adapting P.1121 to fulfil the role. Initial requirements were that the successful aircraft be capable of 600-nautical-mile radius of action at sea level with Mach 1.3 dash carrying a nuclear weapon combined with super-accurate navigation and all-weather radar reconnaissance. This OR, which would eventually surface as GOR.339, would become the most politically tainted programme of the next decade and lead to major upheavals within the aircraft industry but, for now, all that was in the future.

In October 1956 the vexed question of the weapon system concept raised its head, several official sources recommending that Hawker bite the bullet and offer the service a complete weapon system including search radar and missile performance, coupled with ground control to provide a definite defence system. Camm, of course, was opposed to such 'American' concepts and initially resisted such ideas though later he would surprise Hooper by telling him that he was now the weapon-system specialist.

On 15 October 1956, DORA, Air Commodore Kirkpatrick (later, April 1957, Air Vice-Marshal Kirkpatrick) and DOR/B visited Camm to give their views on the P.1121 as:

P.1121 mock-up in Experimental Department at Hawker Aircraft, Kingston upon Thames.

a) a fighter for UK defence,
b) a fighter/bomber and
c) a tactical strike reconnaissance aircraft.

As a fighter, Kirkpatrick did not believe that Treasury would support the P.1121 since two other projects (F.23 and F.177) were already in progress and Hawker's offering did not match up to the requirements of OR.329. As a fighter-bomber, Kirkpatrick was even less enthusiastic, pointing out that no aircraft had ever been specifically ordered against such a role. With regard to the tactical strike reconnaissance role, he told Camm that the OR for this role was at an early stage but preliminary performance requirements included an all-weather twin-crew aircraft capable of radius of action of 600 miles at low level, some of this

P.1121 mock-up with DH Gyron engine in front, probably a space model.

at high subsonic speed and dash of Mach 1.3, maximum runway length of 1,000 feet, tactical nuclear store comprising two 1,000lb and comprehensive navigation suite including sideways-looking radar. This requirement would later be issued as GOR.339. However, in a later meeting in November 1956 between Air Ministry and Air Staff officials, the use of P.1121 in the TSR role was dismissed, the members present believing that 'the aircraft fell too far short of our requirements for a Canberra replacement to be considered in that role'.[11]

By the end of the year the design had matured with a slight increase in length to 66.5 feet. Engine choice was now proposed to include the Rolls-Royce

P.1121 centre and front fuselage under construction with fuselage space model behind.

Conway and Bristol Olympus units while armament was predicted to include a 'target marker' (i.e., nuclear store) on the inboard wing station balanced with a 300-gallon drop tank on the other wing. Maximum fuel carriage had increased to 1,500 gallons internal and 1,000 gallons externally though this meant that no other underwing stores could be carried. January 1958 would see the last single-seat version offered; thereafter design would centre around a twin-seat configuration including a developed version for the Royal Navy featuring folding wings and nose and side- by-side crew complement.

In January 1957 Hawker was visited by Air Commodore McGuire, Group Captain Johnson and Wing Commander Jackson of DOR/Tac branch at the Air Ministry who were briefed on the P.1121 concept and agreed that the aircraft should fulfil the sort of role embodied in the new GOR.339 for a Canberra replacement. At the end of January, with a change of prime minister (Sir Anthony Eden replaced by Harold Macmillan), large cuts in defence expenditure were announced. As the day diary noted, 'The new Defence Minister (Mr Duncan Sandys) visited the US to discuss possible purchase of guided weapons etc., and a cold chill is being felt throughout the British military aviation industry. The public press is expressing views of the fighter being obsolescent'. In other words, the public were being softened up for the sweeping reductions to the country's aircraft programmes planned for the April review.[12]

In the meantime, work continued to progress the P.1121 project, wind-tunnel testing of scale models and air intake designs proceeding apace. In February, in a foretaste of what was around the corner, telephone contact with AVM Kirkpatrick (DCAS/OR), Air Commodore McGuire (OR), and Air Commodore Bell, Head of Airborne Radar (MoS), to arrange detailed discussions on various electronic aspects relating to the new OR were unhelpful: 'The results were most disappointing in that neither Air Staff nor MoS were willing to speak, apparently being afraid to commit themselves. We were informed, however, that the OR would be official in March and four selected firms would be officially acquainted of its contents.'[13]

General view of Experimental Department showing P.1121 centre and front fuselage sections now mated.

March 1957 brought further gloom to the proceedings. Camm visited DCAS Sir Geoffrey Tuttle whom he found to be 'very depressed regarding the future of manned aircraft in the RAF', confirming that Avro's new design, the Avro 730 Mach 3 bomber, had been cancelled. On 19 March 1957, Air Vice-Marshal Kirkpatrick, Air Commodore Roulston and Wing Commander Lowe of DOR visited Kingston to discuss P.1121 and view the mock-up. They appeared uninterested in any defence roles for the aircraft and considered that the aircraft did not sufficiently meet the requirements of the forthcoming OR. This was hardly surprising since Roulston had given vent to his feelings about the project the previous year. 'I am amazed that we are giving it [P.1121] second thoughts, even as a fair weather aircraft. Hawkers have apparently not yet grasped the significance of the weapon system concept. … as a weapon system it would be useless.'[14]

In fact, by this time Air Staff thinking had already rejected P.1121 as it then stood. Air Marshal Tuttle, in a letter to Air Marshal Sir Claude Pelley spelled out in February 1957 that, 'All our examinations of the Hawker proposal to date have amounted to a rejection of the aircraft as an interceptor because it falls far short of the performance required of the fighter to OR.329. In the role of a Canberra replacement, the Hawker proposal falls far short of our provisional requirements for a tactical bombardment/reconnaissance aircraft.'

However, 'until we are certain what projects are going to remain in the programme after the defence cuts currently under discussion, I believe that we would be wrong to discourage Hawker's to continue for the time being with their private venture. What it amounts to, therefore, is that there is no official Air Staff interest at this moment, but, because of the absence of decisions on other projects, I believe that we should continue to give Hawkers what advice we can short of implying official interest.'[15]

In May, after the announcement of the defence cuts, a fairly comprehensive assessment of the design was produced, prepared for the Air Staff by OR.24 at the Ministry of Supply. For the offence role, it noted that the latest Hawker brochure, received on 10 March 1957, proposed a single-seat aircraft powered by a Rolls-Royce Conway Co.11R giving a cruising speed of Mach 0.85 at altitude and a dash speed of Mach 0.9 without reheat, or Mach 1.05 low level and Mach 2.1 at high altitude with reheat. Normal take-off weight would be 43,700lb with 1,500lb fuel and nuclear store carried externally, rising to 48,200lb all-up weight. Take-off would be achieved at just under 2,000 yards and landing 1,450 yards in normal conditions. Navigation would be by Yellow Lemon, Doppler computed to give range and bearing. The Ministry assessment was critical of

Close-up of P.1121 fuselage with space model behind. Note the recesses for the pop-out rocket packs.

the navigational arrangements, in particular the requirement for the pilot to fly and navigate using Yellow Lemon, the limitations of a Doppler system being highlighted. Concern regarding the external carriage of the nuclear store were also expressed, in terms of temperature rise at high speed.[16]

Total R&D costs were expected to be £9 million, excluding engine development, comprising £6 million for eight development aircraft and £3 million for design, flight test, jigs, tools and spares. In conclusion, the report author stated: 'Because it is a single-seat aircraft which carries stores externally and because of its inadequate radius of action, the P.1121 is unacceptable as a strike/reconnaissance aircraft for the Royal Air Force. The firm implicitly hold this opinion themselves by referring in the brochure to a two-seat development fitted with additional navigational aids, including forward and sideways looking radars. This development … cannot therefore be considered as an interim tactical strike/reconnaissance aircraft but could be examined against GOR.339 in company with the projects of such firms as are invited to contribute.'[17]

Close-up of P.1121 front fuselage showing the ventral air intake location and wing attachment points.

An assessment of the defensive role, i.e., as an interceptor, also in May 1957, was similarly dismissive, Wing Commander Willder noting that 'there is insufficient provision for pilot navigation … Yellow Lemon as it stands does not have a suitable presentation for our type of operation'. The lack of search radar was also seen as a serious omission.[18]

With the cancellation of most fighter projects following the 1957 Defence Review, announced on 4 April, the industry was left with little in the way of work to occupy their staff and redundancies were widespread. Really, the only project left was GOR.339 and so every company with any pretence to military aircraft design, turned to the operational requirement; it was really the only game in town. GOR.339 included strike over a 1,000-mile radius, comprising subsonic high-altitude cruise, supersonic high-altitude cruise and low-level transonic profile to deliver a nuclear weapon with great accuracy in all weathers with an all-up weight in the order of 100,000lb AND a short field performance – a tall order indeed.[19]

Hawker, of course, although concerned at the way the wind was blowing, did not feel immediately threatened by the Sandys White Paper strictures, since P.1121 was now being offered as a strike aircraft rather than an interceptor. However, in response to certain Air Staff elements who disliked single-engined aircraft, Ron Williams schemed a twin-engined version at the end of May 1957, around two Rolls-Royce RB.133 engines, a supersonic development of the Rolls-Royce RA.24 Avon and designated P.1125. The engine offered performance of Mach 1.3 at sea level and Mach 2.4 at high altitude, delivering 13,000lb thrust dry and 18,150lb with reheat.

The following month, in an effort to get to grips with the weapon system concept, still something of a bête noire for Camm who was by now well outside his comfort zone, Project Office staff decamped en masse to RRE Malvern for a seminar on future navigation systems pertinent to the low-level strike role of GOR.339. As Bob Marsh noted in the day diary, 'The equipment proposed was very elaborate and of vast conception'.[20]

In 1957 Brian Buss was engaged as a section leader in the Experimental Drawing Office to work on the P.1121 fin design. His time there was to throw up a few surprises as various teams worked on the detailed design of the aircraft. 'Jack [Simmonds] was responsible for the cockpit and forward fuselage design. In this respect I was astounded to see articles and photos being used as a design guide depicting cockpit details of Convair's two fighters, the F-102 Delta Dagger and the F-106 Delta Dart, both of which were flying at the time.'

Buss's concern was subsequently aroused by the lack of space in the rear fuselage to accommodate the frames between the Gyron reheat jet pipe and aircraft skin: 'the frame depth could barely exceed four inches.' His concern was that his fin spars would have to attach to these frames. The sourcing of material for these had not been without difficulty.

'To a large extent, the basic design of the fin was determined for me. Production of high tensile steel in the UK was at that time limited and directed to military projects on a priority basis. As the P.1121 was a private venture project it had to seek supplies of this type of steel elsewhere. It found a Belgian supplier, but in order to have the material when construction commenced it had to place the order some 15 months before. Hence someone had to estimate the size and shape of the various forgings well before the fin was designed in any degree of detail. A forging was required to ensure the grain of the material flowed in the correct directions to gain maximum strength and avoid the onset of cracks etc. So one day I was presented with drawings of three huge steel forgings which I had to incorporate in my design. Digger Fairey had made the initial estimate to place the order and, although difficulties arose, a design evolved in which all could be used.'

In his short time with Hawker, Buss managed to cross swords with the 'Great Man' himself.

'During my second week he [Sir Sydney Camm] came in and leaned on my board and asked how I was doing. Immediately both Frank Cross [Head of Experimental DO] and Harold Tuffen [Assistant Chief Draughtsman] sprang out of the office they shared and were beside us. Sir Sydney looked at my layout consisting of a two-spar design dictated by the steel forgings and went into quite a rage. This was along the lines of what was he paying me all this money for if I could only come up with a traditional design? I tried to point out the restrictions placed on the layout and the time lag that would result if new steel members were to be ordered at this stage, but he would not listen. I said that I could easily produce a multi-spar design similar to the P.1121's wing but the rest of the aircraft would be made and waiting another year for its fin. Again, he would not listen and all the time my two superiors stood there not saying a word.

P.1121 full-sized model of air intake for use in engine testing.

The Search for a Supersonic Fighter   93

P.1121 wing under construction showing the four-spar arrangement.

'At that point I left them to obtain a roll of paper from the drawing stores on the other side of the DO in order to present another layout for Sir Sydney's approval. Little did I know that Sir Sydney thought that I had walked off and ignored him. We met face to face in the centre of the DO when I was returning to my board. He was beside himself, shouting "What the bloody hell do you think you are doing?" With our noses almost touching I shouted, "If you will only get out of my bloody way, I will do what you bloody well asked for." At that he stormed off and every draughtsman in the DO had their heads well down.'[21]

Needless to say, nothing further happened after this contretemps. Camm was perfectly affable to Buss who had passed Sir Sydney's little test more than adequately.

As work continued, July brought more depressing news to the Kingston team. A visit to Air Marshal Sir Geoffrey Tuttle DCAS by Camm resulted in the prescient revelation that he (Tuttle) could give little official support to Hawker's aircraft, 'mainly on account of the financial situation. He also confirmed that the full GOR.339 aircraft would, in all probability, be abandoned.' De Havilland, the engine manufacturer of choice for the project, now added to the gloom when testing of the Gyron began behind the full-scale intake supplied by Hawker, compressor surging occurring well before full RPM, preventing the engine attaining full thrust. The intake now began a somewhat peripatetic lifestyle, shuttling back and forth between Leavesden and Kingston as various remedies were tried in an effort to alleviate the surging, with little or no improvement. This would lead to renewed interest in the Bristol Olympus engine as an alternative to the Gyron and the latter would immediately prove its worth by achieving full throttle behind the intake. It appeared that the Gyron was designed to run much too near to its surge margin, a problem absent on the Bristol engine.[22]

In September 1957 a board meeting agreed that although P.1121 work would continue it would be at a much-reduced rate of expenditure, 20 per cent of the previous rate. However, unbeknown to the Kingston team, while the official GOR.339 had been distributed to industry in September, Hawker had not received a copy. Probing by Camm revealed behind-the-scenes manoeuvring at Hawker Siddeley Group HQ where the decision had been taken to proceed with just one submission from the group, and it would not be Hawker's, Avro being nominated for the task, perhaps as compensation for the cancellation of the Avro 730 project. Camm made his feelings known in no uncertain terms and received grudging acknowledgement that each site could respond to the OR.[23]

Whilst the single-seat P.1121 design proceeded through its various milestones toward an actual aircraft prototype, in the Project Office work continued on alternative layouts. Interest had by now turned to a two-seat design to bring the project more into line with GOR.339 – powered by a Bristol Olympus engine – reflecting the requirement for a strike fighter to be capable of reconnaissance duties but dropping the interceptor role. The second crew member would take on a navigation role, using the multiple sensors including radar and its attendant CRT scope, Doppler, UHF Homing (Violet Picture) and ILS. Length was now 67 feet 8 inches and span 37 feet. Flight refuelling was possible via a probe mounted to port adjacent to the cockpit. Carriage of a target marker was now on a centreline pylon, freeing up the wing stations for external fuel equating to a further 1,000 gallons in drop tanks while the pop-out rocket packs had been deleted, the space being used for the increasingly complex electrical system required to drive the various sensors. The engine was now specified as the Bristol Olympus 21R and a tail parachute was added to enhance short-field performance.

'Stage B' of this design, dated November 1958, would have seen length increase again to 70 feet and span to 39 feet, available wing area increasing from 474 to 509 square feet; the engine was now proposed as the Bristol Olympus 22R

The Search for a Supersonic Fighter 95

P.1121 cockpit mock-up with radar display at top right.

with an internal fuel capacity of 1,600 gallons. Navigation equipment continued to expand, now including sideways-looking radar coupled to a rapid processing unit. Despite Camm's best efforts to sell the idea to the powers that be within the RAF and the Air Ministry, there appeared to be no interest in the design, to a great extent because the winds of political change, blowing ever stronger since the Sandys White Paper, meant that the Air Ministry had their eyes on a far juicier acquisition which had been promulgated in GOR.339. Notwithstanding Hawker's preference for the Olympus, given the problems with the Gyron engine, a watching brief was maintained on flight trials of the latter engine. On 17 February 1958 Duncan Simpson, one of the Hawker pilots, flew with the de Havilland test pilot from Hatfield in the Short Sperrin VK158 flying test-bed, the lower position in each nacelle being occupied by de Havilland Gyron engines, of which the port engine was instrumented. Performance measurements were undertaken up to 38,500 feet including relight trials, all being completed satisfactorily.[24]

P.1129

By the beginning of February 1959 P.1121's journey had effectively ended with no concrete interest being shown in the aircraft. By now, the prototype was half-built, most of the fuselage being complete and wing manufacture under way. The incomplete aircraft would find its way to the collection of airframes held at RAF Cranwell as a teaching aid. Henceforth, work continued in a further attempt to fulfil the requirements of GOR.339 under the new designation P.1129, first designated in November 1957, Ron Williams drawing an enlarged version of his earlier P.1125 but with greater wingspan. Overall length was now 72 feet nine inches, span 48 feet eight inches with a fuel capacity of 2,250 gallons internally and a further 1,600 gallons externally. Two seats and two Bristol or Rolls-Royce engines with hemispherical shock cones in the intakes made for a large aircraft more suited to the long-range role envisaged. This was then refined with an increase in length to 73.5 feet and alterations to the intakes, variable ramps replacing the shock cones. Fuel carriage was also increased to 3,000 gallons internally.

By January 1958 it was apparent that the Avro and Hawker designs were similar, agreement being obtained that both companies could submit their proposals to the OR branch, though the Hawker Siddeley Group board would make the Air Ministry aware that the Avro 739 submission was their preferred choice, in effect telling OR to ignore the Hawker submission. With the submissions delivered, Hawker continued with the project whist various interested parties within the Air Ministry and government research sites pored over the submissions.

By July, with little information leaking out as to progress with the Air Ministry, concern within the Hawker Siddeley Group board regarding the multiple submissions was growing, culminating in a boardroom coup that saw Sir Frank Spriggs (seen as biased towards Hawker) ousted and replaced by Sir Roy Dobson, the Avro Chairman. A decision was now taken to resubmit a single project to the OR, driven in large part by Dobson. Following a fraught visit to Avro, Manchester by Camm, Williams and Hooper, a decision was reached which would see the P.1129 form the basis of the submission but with certain features taken from Avro's project added; the project being under Camm's rather than Avro's control, this being submitted in late 1958. The revised design had deleted the variable ramp intakes, these being reworked with Avro's forward swept high mounted intakes and blown leading and trailing edge flaps together with differentially operated tailplane for high-speed cruise.[25]

In January 1959 it was announced that the GOR.339 competition had been won by Vickers, supported by English Electric. Further, it became clear later that the decision to award to Vickers had been taken in the summer of 1958, thus making all the work carried out by Hawker Siddeley Group superfluous. Of course, Hawker could not know that their failure to win the TSR.2 project would be to their benefit, though this would become clear as time went on and TSR.2 would become the most politically tainted UK aircraft programme of the twentieth century.

Reasons for the failure of the P.1103/P.1121 and P.1129 designs are many and various. Of the various submissions to F.155T, only the Hawker design was realistically capable of entering service in the timescale specified. This was achieved by using a fairly conventional design capable of high performance around Mach 2. That it did not follow the letter of the specification was a deliberate stance to enable production in a timely manner but this same 'failure' riled the various Air Ministry officers who expected 'their' specification to be followed rigorously. There appeared to be irritation among certain civil servants with Camm's manoeuvrings at high level seeming to undermine their own positions. Camm's aphorism 'Follow the spec and you are dead', neatly encapsulated the Hawker design ethos: don't follow the OR religiously; its authors will have a limited understanding of what is possible. Rather, use the OR as a basis for a design that can be produced to answer most of the requirement at a cost and within a timescale that is realistic.[26]

With the F.155T tender lost, the decision to continue with a strike version was the correct choice since any further offering as an interceptor would have come to grief with the Sandys White Paper. Unfortunately for Hawker, P.1121 was being offered to an Air Ministry trying to re-align itself and its requirements with a missile-dominated future, or so it seemed at the time. The emergence of GOR.339 once again offered a way in which Hawker might still get their project into service, though now as a low-level bomber. While the resulting P.1125 and P.1129 were undoubtedly capable designs, politics, both at national and board level, ensured that neither would ever be built. Nationally, the lure of the TSR.2 project had concentrated the minds at the Air Ministry and the OR branch. Since this was to be the only major project available to them, everything, bar the kitchen sink, would be specified for the strike aircraft. Hawker's offering, essentially 'TSR.2 lite', while less expensive and likely ready sooner, was overlooked for the 'sexier' BAC offering. The resulting over-complication and soaring cost would not be helped by the forced marriage of Vickers and English

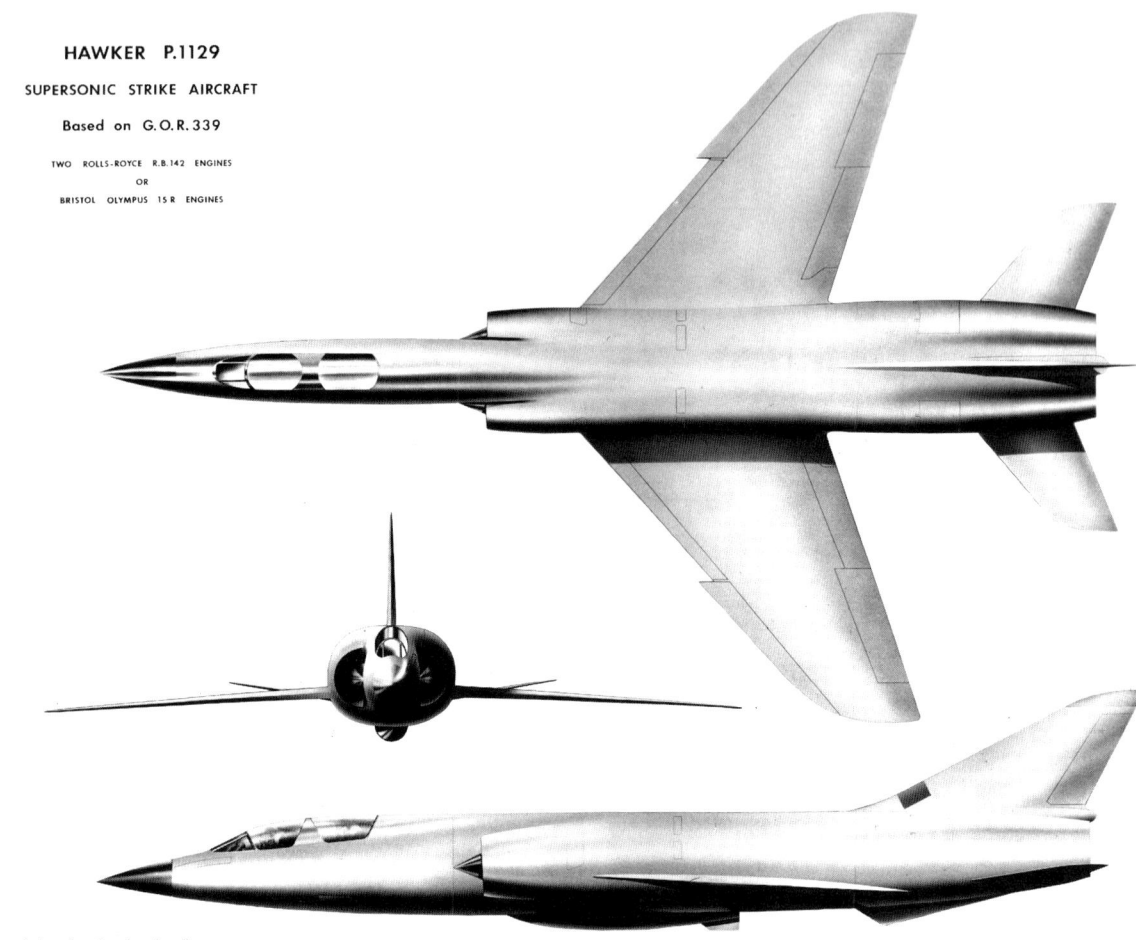

P.1129 three view showing Kingston's submission prior to changes incorporating aspects of the Avro design.

Electric into BAC, as two very different management teams sought to impose their will on a project mired in civil service interference. At Hawker Siddeley board level, the uncertainty over how the group should respond to GOR.339 caused ill-feeling within it and undoubtedly weakened Hawker's project in the eyes of the Air Ministry. Submitting two tenders instead of one was a sure way to irritate the civil servants further.

Would TSR.2 have been cancelled if P.1129 had been chosen? Possibly not, since it could have been flying sooner and thereby been further along the development road when decisions came to be made. Whether it would have suffered the same fate as the BAC submission (endless additions and management by committee) is uncertain and one of the many questions that will remain unanswered.

Could P.1121 have been an effective tactical strike aircraft? Almost certainly. In the US the Republic F-105 Thunderchief appeared slightly earlier than P.1121 would have entered service and had many similarities to the Hawker design. In offering a similar aircraft to fulfil the tactical/strike role, Republic were successful in launching the programme, eventually building 830 aircraft, the Thunderchief seeing extensive and successful service in Vietnam before being phased out, the last aircraft being retired in 1983. The P.1121 would certainly have found a useful role in the northern European theatre but ultimately defence funding and political will were simply not available. Richard Worcester, writing in 1966, found the decision to ignore P.1121 puzzling. 'It was started in 1956, some three years after the F-105, and for the Government to have declined it, waited three years and then started TSR.2 which they should have known was too expensive, too large and too late, is another example of poor decision making abilities.'

As a postscript, it is of interest that prior to the successful partnership of Vickers and English Electric, the latter had carried out considerable work with Short Brothers and Harland in an effort to answer the extended range requirement of GOR.339. The solution was envisaged as a composite aircraft comprising the EE P.17A strike bomber which would be launched and recovered by the Short P.17D VTOL lifting platform, a delta planform powered by no fewer than fifty-six Rolls-Royce RB.108 lift engines. Whether this was known to Hawker project staff is unknown but initial study by Chris Hansford examined a similar lifting platform for the P.1129, this time powered by the new (October 1958) Bristol BE.53 engine. Two layouts were examined, the first comprising a twin fuselage with P.1129 carried between them, powered by ten BE.53 lift engines and the other, a single fuselage with enlarged wings under which one or two aircraft could be accommodated, powered by twenty BE.53 engines. While a sketch exists, no proper drawings appear to have been produced.[27]

Chapter 5

Going Up: V/STOL Studies

Perhaps more than any other concept, vertical take-off and landing – V/STOL – defined the work of Hawker Siddeley Aviation in the minds of the public; the Harrier becoming *the* V/STOL aircraft of the later twentieth century. Seen by some, including some at Hawker itself, as something of a flash in the pan, STOVL (as it is now known) is now 63 years old and just getting into its stride with the entry into service of the F-35B Lightning II, itself expected to remain in service for the next thirty years. VTO has been presented by some as the product of mere idle doodling by a bored designer one day at Kingston, the idea arriving out of thin air, to design an aircraft that did not need a long runway. The truth is rather more prosaic and protracted.

With the advent of the Cold War, following as it did the murderous years of the Second World War, an abiding concern of the military planners of the western nations was that of being taken by surprise by Warsaw Pact forces, of being caught with one's air forces on the ground where they were vulnerable to destruction before they could enter the fray. Recent history had shown that it was not even necessary to destroy the aircraft; one might merely crater the exposed runways, thus neutralising the air offensive capability of the enemy by stranding them on the ground. NATO military doctrine therefore evolved in the 1950s to ensure that in times of tension, air assets might be capable of dispersal to other airfields or even to locations which did not require an airfield at all. For such an aircraft, vertical take-off and landing (VTOL) capability would be indispensable.

Research towards evolving such an aircraft was alive in various western nations in the early 1950s. In the UK, NGTE at Pyestock had carried out research into early principles likely to relate to VTOL aircraft. The results, published in September 1953 as 'Powerplant Requirements for Vertical Take-off Aeroplanes using Jet Lift', examined the theoretical requirements for VTOL flight, a subsequent contract being allocated to Rolls-Royce to develop a vehicle – the Thrust Measuring Rig – to evolve a suitable reaction-control system to allow control in hovering flight. While Rolls-Royce's TMR had demonstrated control and three-dimensional manoeuvre in the hover via jet reaction controls at the vehicle's extremities, another RAE contract, this time allocated to Short Brothers and Harland of Belfast, sought to develop a research aircraft with which to test both lift jets and the reaction-control system used in the TMR assisted by a powerful three-axis auto-stabilisation system, considered essential to allow accurate control in the hover.[1]

From this, Short Brothers had developed an aircraft – the Short SC.1 – capable of VTOL, courtesy of four lift engines under the vehicle to provide jet lift and another at the rear to provide horizontal flight. The engine – the RB.108 – had been developed by Rolls-Royce with an excellent thrust-to-weight ratio (8:1) and was extremely compact; as far as Rolls-Royce and government research centres were concerned, this was the answer to the requirement for an engine to lift an aircraft vertically from the ground and its pursuit thus became almost government policy. Excellent maybe, but the lift engine – or rather, engines – since multiples of engines would need to be installed to achieve the required thrust to overcome gravity, imposed a significant weight penalty and space constraint in horizontal flight for no gain; the lift engines were effectively useless for normal flight unless their thrust could be directed to the rear. As Stanley Hooker put it in a note to Camm, 'I should need a lot of convincing that there is any advantage in giving the take-off and landing engines a free ride around the countryside.' This solution then, was cumbersome. What was needed was an engine capable of directing its thrust vector down for take-off and landing and rearwards for normal wing-borne flight.

In the meantime, with the hiatus caused by the failure of the P.1121 discussed in the previous chapter, the Hawker Project Office had examined two possible VTO designs that might just form the basis for a small strike aircraft. In June 1957 Ron Williams examined a possible strike aircraft – P.1126 – drawn around a bank of twelve Rolls-Royce lift engines and two Bristol Orpheus cruise engines. A length of 53 feet and delta wingspan of 32 feet with compound sweep on the leading edge was coupled to banks of six RB.108 engines in each wing root that folded down for operation into a vertical orientation and tucked away when not in use. Conventional leading edge wing-root intakes fed the Orpheus engines for cruise flight. The design offered no STO facility and was therefore equipped with skid rather than wheeled undercarriage.

P.1127

Ralph Hooper had chance to pick up the first brochure produced by Bristol Engines detailing their BE.53 engine that had the capability of vectoring or turning the compressor air through 90 degrees to exhaust either rearwards for cruise flight or down for vertical thrust. Bristol had picked up the concept of vectoring the engine's thrust from a design by French engineer Michel Wibault and simplified his ideas to produce a workable engine. Hooper began to scheme in July 1957 under project designation P.1127 and produced an initial design for something akin to a battlefield liaison aircraft with STO rather than VTO capability. Following discussions with engineers at Bristol Engines, this design quickly grew into an aircraft able to direct the totality of its exhaust down for vertical flight, rearwards for cruise flight or, indeed, anywhere in between. Thus was born the P.1127 which would mature after many years and vicissitudes into the Harrier V/STOL aircraft. It is not intended here to describe in detail the evolution of the Harrier, a story retold many times by other authors, not least by Roy Braybrook and Frank Mason, both refugees from the Hawker Project Office, but to chart the convoluted course of Hawker Siddeley Aviation, which subsumed Hawker Aircraft Ltd in 1963, in their attempts to sell to an ungrateful world a supersonic V/STOL strike fighter.

The P.1127 would fly, or rather hover, for the first time in October 1960 at Dunsfold and would follow an evolutionary course ultimately leading to the Harrier family of aircraft. As seen in the story of the P.1121, supersonic speed was considered of crucial importance by the services for any new military strike fighter. Although it would soon demonstrate supersonic speed in a shallow dive, P.1127 was a subsonic design and therefore immediately dropped down the 'nice-to-have' list at the Air Ministry. What was needed, went the thinking, was a VTO aircraft with the speed and load-carrying capability of the McDonnell F-4 Phantom and so early studies included means of achieving a supersonic V/STOL design even before P.1127 took to the air.

In April 1958 Ron Williams approached this requirement with his P.1132 design for a 'V/STOL Strike Aircraft for Land/Sea Tactical Duties'. The design featured an engine complement of two BS.53 lift/thrust engines with vectoring nozzles and a further Napier Double Scorpion rocket motor in the tail. The wingspan of 37 feet could be folded down to 26 feet for carrier operations, the substantial weight of the aircraft being carried by a robust bicycle undercarriage of quadruple main wheels and twin nose wheel. A refinement of this offered a twin-boom rear fuselage mounting a high tail and the undercarriage reverted to standard tricycle layout.

P.1136 and P.1137 returned to the lift engine concept, John Fozard drawing a 'Subsonic VTO Strike Fighter' in April 1959, powered by four lift and one cruise engine while Ron Williams schemed a 'Supersonic V/STOL Tactical Aircraft' that featured RB.153 engines, three behind the cockpit, one on each wingtip capable of swivelling through 90 degrees and a further two in the rear fuselage with a clang box to allow either cruise or lift power. Williams offered an alternative layout that brought the wingtip and aft engines together into underwing nacelles capable of rotating through 90 degrees from vertical to horizontal, as well as the dedicated lift engines in the fuselage.

Not to be outdone, in September 1959 Roy Braybrook offered a Mach 3 futuristic concept VTOL interceptor for the Royal Navy fitted with six RB.153 lift engines and a further three RB.153R engines for cruise flight. The P.1138 tailless design featured innovative canard control of pitch and forward-swept intakes mounted above the wing-root trailing edge towards the rear of the aircraft. Other combinations of lift and cruise engines included the subsonic P.1139 (two RB.153 lift engines and one RB.163 lift/cruise engine), the supersonic P.1140 of March 1960 (three RB.153 lift engines and one RB.163-1 clang box lift/cruise engine with reheat), both by Fozard, and the subsonic P.1143-1 (six lift engines and four lift/cruise engines) of July 1960 by Braybrook. As will now be obvious, the Hawker Project Office was by no means wedded to the Bristol Engines vectored-thrust engine design, particularly since Sir Sydney had a close relationship with Ernest Hives, Chairman of Rolls-Royce and would use their engines if possible. However, the fact remained that, despite all of the investigation of dedicated lift engines, it was Bristol's vectored thrust engine that proved to be the most economical in terms of weight, thrust, fuel consumption and space and therefore it was this design that would henceforth govern Hawker Siddeley's search for a supersonic V/STOL aircraft.

In the second half of 1960 John Fozard left the Project Office to join the Advanced Projects Group, a Hawker Siddeley Group department (co-located at Kingston) aimed at developing group-wide future projects and was therefore absent while Hooper worked on various V/STOL designs. However, Fozard fell out with his bosses in 1961 and, rather than lose him, they offered him a position in the Project Office in February as No.2 to the Head of Projects, Bob Marsh. He was replaced at APG by Ron Williams, who also returned after a year or so, having fallen foul of Camm in the meantime.

104   Hawker's Secret Projects

In January 1961, with the P.1127 slowly expanding its hovering envelope over the grid at Dunsfold Aerodrome, Ralph Hooper had been attempting to get the P.1127 design to meet the RAF Operational Requirement, OR.345, supposedly written around the aircraft. This was for an aircraft to replace the Hunter in the close-support role that could operate without the need for long runways. Frustrated by the inability to achieve this with the most powerful Pegasus then available, Hooper had lost faith in the OR and returned to the vectored-thrust concept to push forward with a second generation V/STOL strike fighter design. In truth, the lukewarm interest being shown by the RAF in an operational P.1127, once they realised the limitations of the early engine, had convinced Hooper that a more capable aircraft would be required as the minimum to satisfy service chiefs. On 12 January 1961 George Marchant of

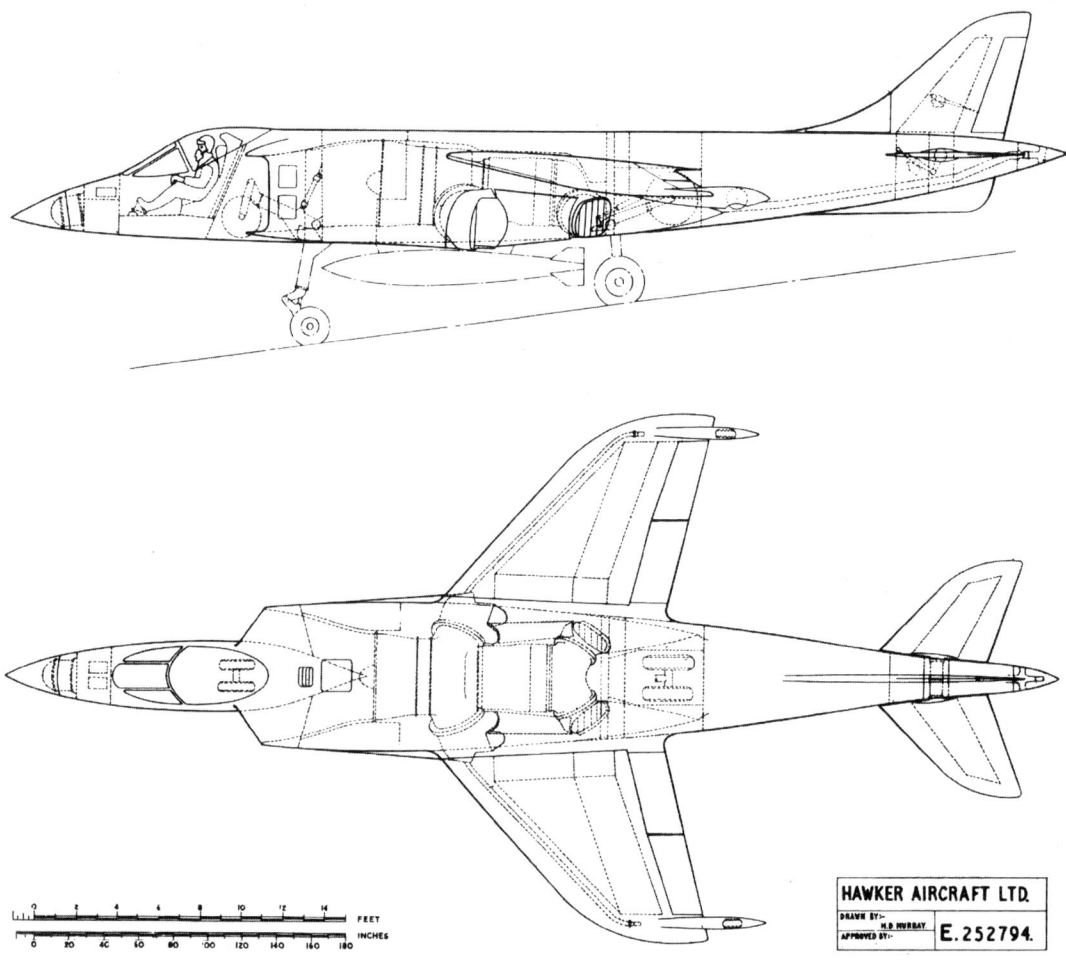

P.1150 GA, supersonic V/STOL strike aircraft.

Going Up: V/STOL Studies   105

Bristol Engines had visited Kingston with news of a larger BS.53 engine which would suit supersonic flight thanks to the theoretical addition of reheat in the cold nozzles. This was just what Hooper needed to allow a supersonic aircraft to be drawn around it.[2]

P.1150

Hooper now began work on design of a supersonic V/STOL aircraft. His first design, P.1150, drawn on 28 January 1961, was based on the planform of P.1127 but used a larger engine incorporating a form of reheat. This engine, the BS.53/6, was coupled with plenum-chamber burning, a means of burning

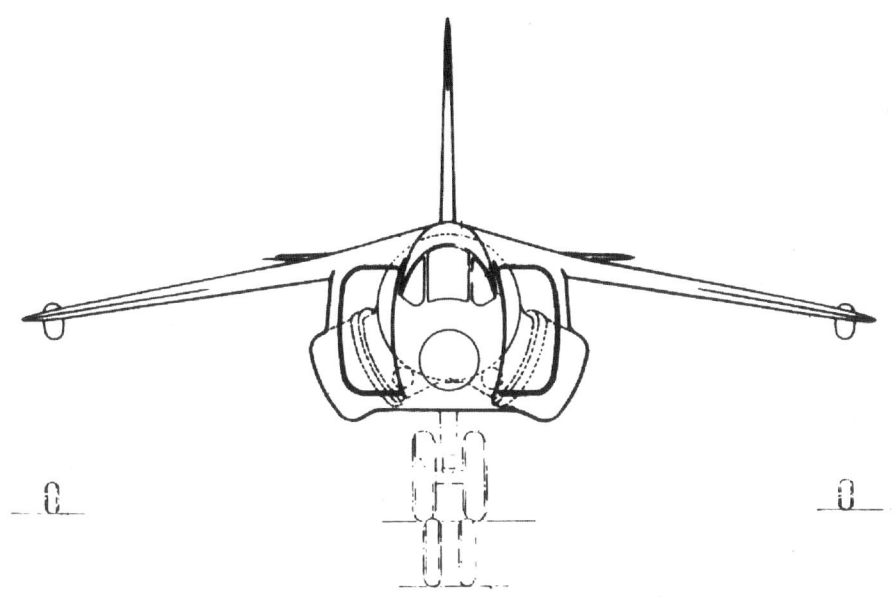

P.1150
SUPERSONIC V.T.O.L. STRIKE AIRCRAFT

SPAN   24 FT. 3 INS.   WING AREA 220 SQ.FT.   WING LE. SWEEP 42·5 DEG.

B/SIDD PEGASUS 5 ENGINE   B/SIDD. PEGASUS 6 ENGINE
   WITH 1200°K P.C.B.   WITH 1200°K P.C.B.
O/A LENGTH 50 FT.   O/A LENGTH 52 FT
INTERNAL FUEL CAPACITY   INTERNAL FUEL CAPACITY
   850 GALLONS   1000 GALLONS

PEGASUS 5 VERSION DRAWN
FRONT AND REAR TANK SECTIONS
EXTENDED FOR PEGASUS 6 VERSION

fuel in the front nozzles, giving an 800°k reheat capability and greatly increased thrust. (The concept had been conceived by Bristol Engines' engineers Peter Orchard and Mike Williams, a patent application being submitted in March 1961.) Length was to be 53 feet and span 26 feet with a leading edge sweep of 42.5 degrees, giving a wing area of 244 square feet. Fuel capacity of 850 gallons offered a useful sortie duration. A major change from the P.1127 design was the drooped and trailed nozzles which provided a less abrupt change of direction for the exhaust at the nozzles and allowed the engine to be removed from below, i.e., without the need to remove the wing as required on P.1127 and, subsequently, the Harrier. Crucially, drooping also gave greater clearance between the jet efflux and the rear fuselage/tailplane, thus reducing fatigue and buffet, two problems that would follow the Harrier through its life.

Detailed work on the PCB capable Bristol Siddeley BS.100 engine had been carried out in response to the AW.681 VTOL transport aircraft in 1962 and was further developed following discussion with a team including Fokker and Republic Aviation as well as with Hawker. The original design, the BS.100/3, offered over 37,000lb thrust using a 56-inch fan which, for Hawker's proposal, was too large. A 0.83 scale version, the BS.100/9 using a 51-inch fan, initially rated at 30,740lb thrust, was later developed into the BS.100/8 which promised 33,640lb thrust for a reheat value of 1,200°k with a slightly larger 52-inch fan and this would be the engine upon which later designs (i.e., P.1154) would hinge.

With Hooper busy on P.1150, Fozard had worked at submitting a proposal to a specification for the Royal Navy which needed to plan for replacement of the Sea Vixen, then just entering service. In June 1961 his design to P.1152 was schemed around yet another multiple-lift jet project, this time with four lift engines and one clang box lift/cruise engine. Kingston's submission was in competition with de Havilland, Blackburn, APG and others.

As an aside to the P.1150 story, Hooper, still working on attempting to answer OR.345, recalled that 'During the summer of 1961 SGH [Stanley Hooker, Chief Engineer at Bristol Siddeley Engines], finally offered an 18,000lb rating Pegasus, and with this … finally met the critical OR.345 sortie. Finally in the early autumn OR.345 was cancelled – as the RAF's response to the P.1127 having shown it could meet the requirement.' In cancelling the OR, the Operational Requirements branch replaced it with OR.356, a far more onerous requirement tailored to a supersonic V/STOL aircraft.[3]

This called for a V/STOL capable aircraft suitable for both land and seaborne use, performing strike/recce and all-weather intercept missions, using a dual weapon system coupled to an AI radar, making a two-man crew mandatory, to be in service with the RAF by 1968 and with the RN by 1969/70. The strike role required the ability to deliver both nuclear and non-nuclear weapons and reconnaissance and air interception by day. Sea level speed not less than Mach 0.92 with a supersonic dash capability was desirable while, for the Royal Navy option, high-level interception would require acceleration from Mach 0.9 to Mach 1.7 at the tropopause with a top speed of no less than Mach 2 and loiter time of at least 2.5 hours. Take-off and landing within 500 feet with a weapon load of 2,000lb was required while a ferry range of not less than 2,500 miles was mandatory.

NBMR-3 Submission

With the P.1127 OR cancelled, it was at this point in the development history of the aircraft that an opportunity to fill a market gap arose in the shape of a NATO requirement for a supersonic VTOL tactical-reconnaissance aircraft. Hawker Siddeley had reason to be confident that their P.1150, suitably modified, would go a long way to meet this requirement, NBMR-3 or AC/169, first received at Kingston on 19 April 1961. Following pressure from the government to seek an agreement with the French and Germans on the P.1127 project, Sandys being keen to reduce development costs, an HSA board meeting had heard that Breguet, Fokker-Republic and Focke-Wulf had been nominated by government ministers as potential associates in the competition, though each group would also be submitting their own proposals. In the event, it was agreed that HSA would co-operate with Focke-Wulf on development of a P.1150 type aircraft with additional lift engines in response to NBMR-3. However, at the first meeting with Focke-Wulf on 6 September 1961 to discuss an agreement the German company was keen to point out that they considered themselves at least equals to HSA and expected this to be reflected in any agreement. Given the lead that Kingston had built up on V/STOL development, this did not augur well for future relations.[4]

The NBMR-3 design, began at Kingston in October 1961, and was soon revised as the P.1150/3 in July to feature Bristol Siddeley's upgraded engine, the BS.100 with 1,200°k PCB and fuel capacity of 1,200 gallons. Length had increased to 55 feet 4 inches and span reduced to 26 feet though still giving 244 square feet.

The designation was now reworked and, to differentiate the refined version from earlier work, in October 1961 the new designation P.1154 was assigned. As

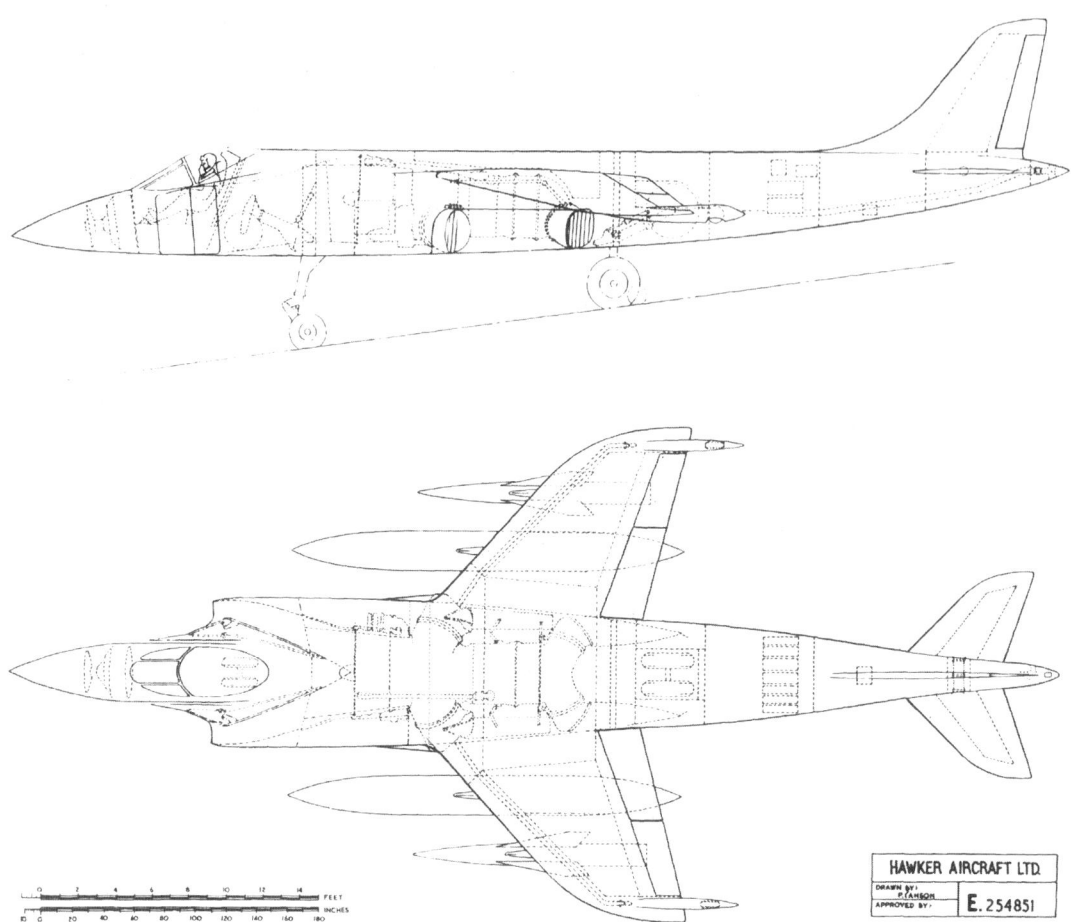

P.1154 GA, single-seat ground-attack, all-weather interceptor aircraft for RAF.

tendered, HSA's aircraft offered a single-seat supersonic V/STOL strike aircraft powered by the Bristol Siddeley BS.100/9 engine with PCB providing 33,150lb thrust and internal fuel of 1,150 gallons giving a combat radius at low level of 295 miles and a ferry range of 2,360 miles. Length was now 55.3 feet, the span and wing area remaining as per P.1150/3. Rectangular intakes with cut-back lips and two supplementary blow-in doors fed the engine which offered speeds up to Mach 2 at altitude with a simple pitot intake and Mach 2.4 with variable wedge intakes. Normal take-off weight would be 28,795lb with a normal weapon load of 2,000lb. Most fuel was to be accommodated in the fuselage due to the thin wing while, behind the main undercarriage, a large equipment bay was available for the avionic kit required to provide inertial navigation and attack system with a lightweight forward-looking radar. Although a large store was

Going Up: V/STOL Studies 109

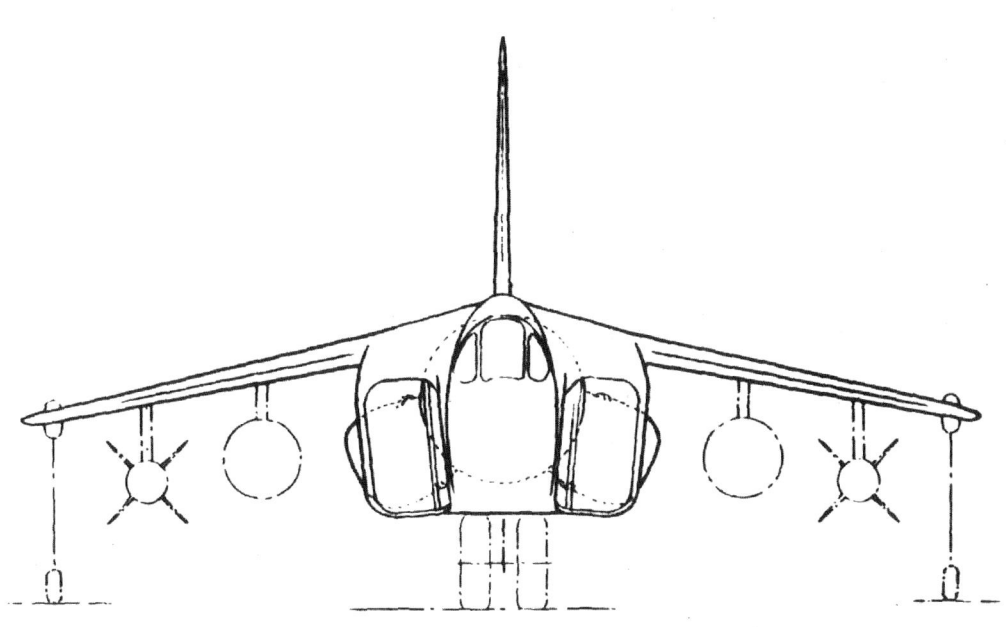

R.A.F. P.1154
SINGLE SEAT GROUND ATTACK / ALL WEATHER
INTERCEPTOR AIRCRAFT

B.S. 100/9 WITH P.C.B. LIFT THRUST TURBOFAN

SPAN   26 FT.   WING AREA   246 SQ.FT.
O/A LENGTH 56 FT. 3 INS.   WING L.E. SWEEP 42-15 DEG.

INTERNAL FUEL CAPACITY 1250 GALLONS

featured on the fuselage centre-line, presumably a nuclear weapon, its ability to survive unscathed in this position with the high temperatures generated during VTO was questionable.[5]

Hawker submitted its NBMR-3 proposal to NATO in January 1962 and awaited results. Rumour favoured the French Mirage IIIV as Hawker Siddeley's main contender, it being offered by Dassault in association with BAC. Slowly, as design and development work proceeded, the HSA design edged ahead of the other proposals by France and Germany and, in July, was eventually declared joint winner of the competition with France's Mirage IIIV entry in a political decision, the French being emphatic that they would go with their own design irrespective of who won, though NATO had no funds to promote construction and it was left to individual NATO nations as to whether they bought either design. Ultimately, it would be left to the RAF and Royal Navy to continue progress with the aircraft, as will shortly be seen. With the abandonment of OR.345, communicated to Camm in January 1962, there had been an expectation – indeed, a Treasury directive – within the RAF that their next aircraft would be the eventual winner of the NBMR-3 and a weather eye had therefore been kept on the progress of the competition. Within the company, three outcomes had been considered:

1) P.1154 wins NBMR-3, political hurdles are overcome and RAF order 120 aircraft. Royal Navy accepts minimum change version and orders 60 navalised aircraft.
2) P.1154 is favourably received by NATO at technical level but political arguments are so protracted that UK Air Staff are forced to proceed unilaterally. Navy's hand now somewhat stronger and press for heavyweight nose radar and second crew member; 200 aircraft ordered.
3) NBMR-3 collapses on political grounds. UK Air Staff proceed unilaterally, if at all. OR.356 pigeon-holed for two years while Air Staff try to justify a national aircraft programme unsupported by the prospect of European sales. Royal Navy now in strong position to insist on two seats etc. RAF version thus considerably compromised.[6]

In the event, outcome 3 was broadly the result. Hooper recalled how, 'On 2nd April 1962, a Royal Navy party led by [Captain Eric] 'Winkle' Brown visited me at Kingston to introduce their "compromised" requirement which they were being politically required to meet with a P.1154 adaptation. It was clear to me that we had offered all we could to NATO and the RAF and had nothing in hand to meet the very different RN requirement. In view of the relatively small numbers the RN would want, I advised that *if we went along this route, we risked losing the lot.*'[7] (Author's italics).

P.1154

The reason for the 'compromised' requirement was that both the RAF and Royal Navy had need to replace existing aircraft with a new design. The RAF needed to replace the Hunter and the Royal Navy the Sea Vixen, and there was political pressure emanating from the Treasury for a common design to be accepted by both parties in the erroneous belief that a) this was possible and b) it would be less expensive than separate designs. The joint requirement for a common aircraft for the RAF and the Royal Navy was promulgated by the then Minister of Defence Harold Watkinson (later Peter Thorneycroft) and ultimately by the Treasury since it was seen as a cost-cutting opportunity. The idea was not new but at that time was current in the military aviation corridors of power in the US, and what was good enough for them was certainly good enough for the UK. Hence was set in train, as Hooper had predicted, the dooming of the entire P.1154 project.

However, as Hooper recalled, 'things were very different from the early days of the P.1127 (where no-one had believed in it in the early days – so that I had things pretty much my own way!), we now had Lickley, Camm, Chaplin and soon Laight all determined to be in on the act and none prepared to say "no". By around the middle of 1962 the Naval Staff Target (NST) to which the P.1152 (and its competitors) responded had been withdrawn thus leaving JWF [Fozard] once again without a job. Nevertheless, he knew about naval V/STOL – didn't he? And Hooper had said the joint RAF/RN couldn't/shouldn't proceed.' His influence over the project waned over the year and finally, in October 1963, John Fozard was made Chief Designer P.1154. As Fozard headed up P.1154 work,

P.1154 RAF desk-top model.

P.1154 Royal Navy desk-top model.

Hooper returned to work on developing the P.1127 into the more advanced Kestrel FGA.1 in anticipation of getting a squadron of development batch aircraft into service under the auspices of the Tripartite Agreement with West Germany, the USA and the RAF.[8]

The operational requirement to which P.1154 would respond was OR.356/AW.406, issued in April 1962 and which called, as seen above, for a common design to suit the requirements of two very different users. The RAF requirement called for an aircraft to replace the close-support and reconnaissance Hunter by 1968, capable of V/STOL, able to provide strike/reconnaissance from dispersed sites in support of the Army. This would broadly reflect the NBMR-3 specification. The Royal Navy required a two-seat, high-altitude interceptor with powerful radar and missile armament; any strike role was strictly secondary. It should be capable of catapult launch from its carriers, the Royal Navy seeing no use for any V/STOL capability.[9]

The Kingston Project Office's first brochure, issued in October 1962, offered an airframe of 55 feet 3 inches and wing-span of 26 feet swept at 42.2 degrees with a t/c ratio of 6 per cent at the root and 5 per cent at the tip, powered by the BS.100/9 offering 31,550lb thrust with PCB and 1,380 gallons of internal fuel, 576 gallons external. By March 1963 this had matured into a submission using the BS.100/8 engine of 32,420lb thrust (phase 1) at normal PCB operation and 35,170lb (phase 2). The RAF version was now 55 feet with a wingspan of 26 feet and an area of 246 square feet. With an internal fuel capacity of 1,100 gallons, normal take-off weight would be 30,170lb. Maximum level speed at altitude would be Mach 1.92 with an endurance on patrol of 1.6 hours. The Royal Navy version would be 58 feet 5 inches long with a wing of 30 feet span, reduced to 22 feet when folded and area of 287 square feet. Internal fuel capacity would be 1,600 gallons and normal take-off weight 38,960lb. Patrol endurance of 2.35 hrs with a maximum level speed of Mach 2.13 at altitude, using semi-cone shock intakes, would give the Royal Navy a formidable aircraft. Weapon loads differed, the RAF version being bombs with secondary load of four Red Top missiles assisted by a lightweight radar suite while the Royal Navy's primary weapon would be four Red Tops and secondary load of 4,000lb bombs with a full-sized airborne interception (AI) radar, both being linked to an inertial navigation/attack system.[10]

As detail work began to refine the response to the two requirements, commonality was seen as potentially around 80 per cent of the total but by May 1963 this had reversed and was now only around 20 per cent commonality, mainly due to the continuing demands for changes by the Royal Navy. The RAF version remained fairly constant but the Royal Navy's aspirations saw a steady divergence in the designs. The Royal Navy requirement was for an interceptor capable of performing Combat Air Patrol (CAP) for several hours at least 100 miles from the carrier. In order to see the enemy bombers early enough, a large radar was required together with a second crew member to operate it. A rapid climb and supersonic level speed were needed to accomplish the interception, thus requiring a much larger fuel fraction than the RAF version.[11]

The requirement for Mach 2-plus performance at altitude necessitated two-shock intakes which, with the large AI radar, second crew member and new catapult-friendly undercarriage resulted in a heavier airframe, which required larger wings. Also, the zero-track bicycle undercarriage would, it was at first held, interfere with catapult-launch facilities and so the main undercarriage was moved from the fuselage to the wings and accommodated in mid-span sponsons. (In fact, with a revised catapult shoe, it was later shown that the bicycle undercarriage was feasible). To improve deck lift capability the nose was designed to swing through 180 degrees to shorten the length, as per the later Sea Harrier design. One might question why the Royal Navy felt the need to retain catapult launch when it was being offered an aircraft capable of short take-off, but the deck behind the catapult on its existing carriers was not free, necessitating alterations to accommodate a launch bridle, strengthened nose leg and hold-back facilities on the aircraft. So much for VTO but what of the ability to land vertically? By this time the design had grown to an extent where the airframe mass had out-run the available engine landing thrust such that vertical landing was no longer possible necessitating the fitting of an arrestor hook and beefed-up main undercarriage.

As well as the airframe, the internal systems required were entirely different; they were after all being asked to perform entirely different roles. Driven by

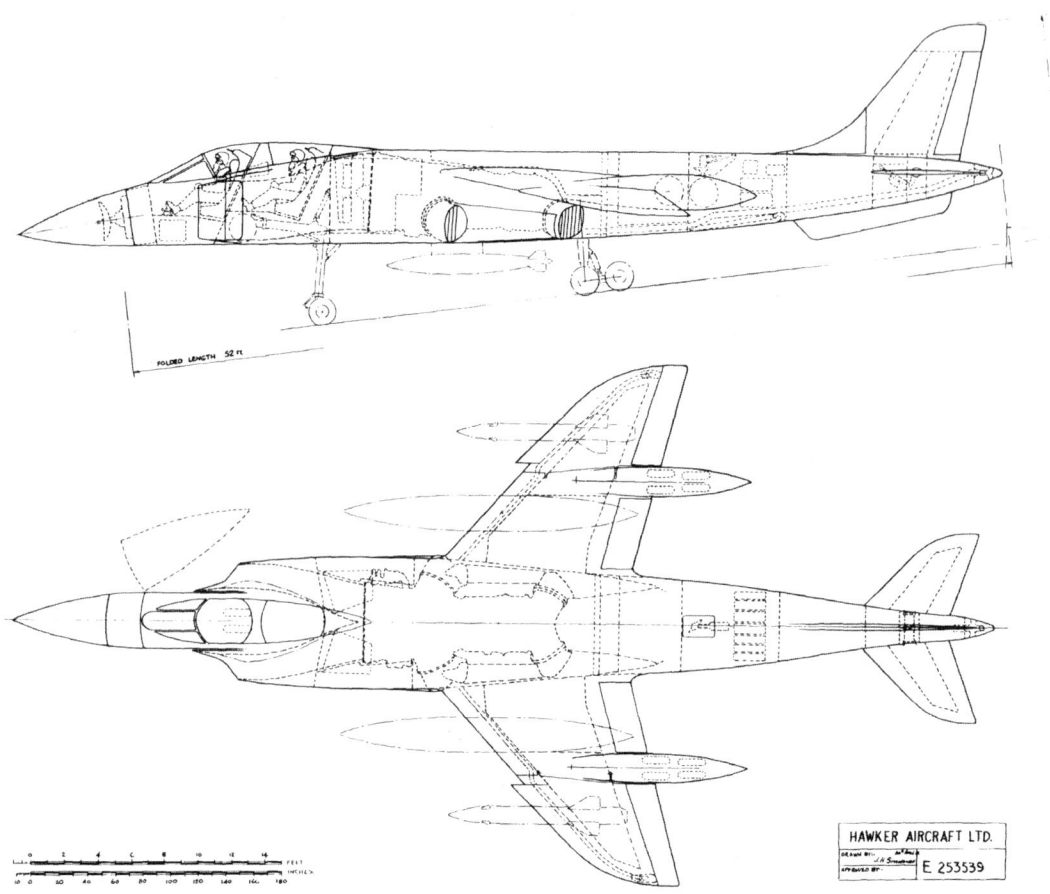

P.1154 GA, two-seat supersonic interceptor, strike aircraft for Royal Navy.

the Royal Navy's stance, by June 1963, the designs were, in effect, two separate aircraft. The Defence Research Policy Committee, in seeking to bring the designs back to a common base, demanded one common design be accepted by both parties and Kingston duly reworked the separate designs into a single aircraft.

Inevitably, in trying to merge two disparate requirements, neither the Air Ministry nor the Admiralty were happy with the resulting design, the Ministry of Defence rejecting HSA's proposals. In June 1963 Mr Nicholson, Director General of Scientific Research (Air) revealed that the MoA wished the company to consider two proposals as the basis for a truly common design. P.1154/1 (a single-seat P.1154 RN with the large wing and soft tyre tricycle undercarriage) and P.1154/2 (a navalised P.1154 RAF with bicycle undercarriage and with some AI capability without catapult facilities). Kingston's investigation showed

SABA artwork showing aircraft in formation with V-22 Osprey tiltrotors.

P.1121 artwork showing the aircraft at low level.

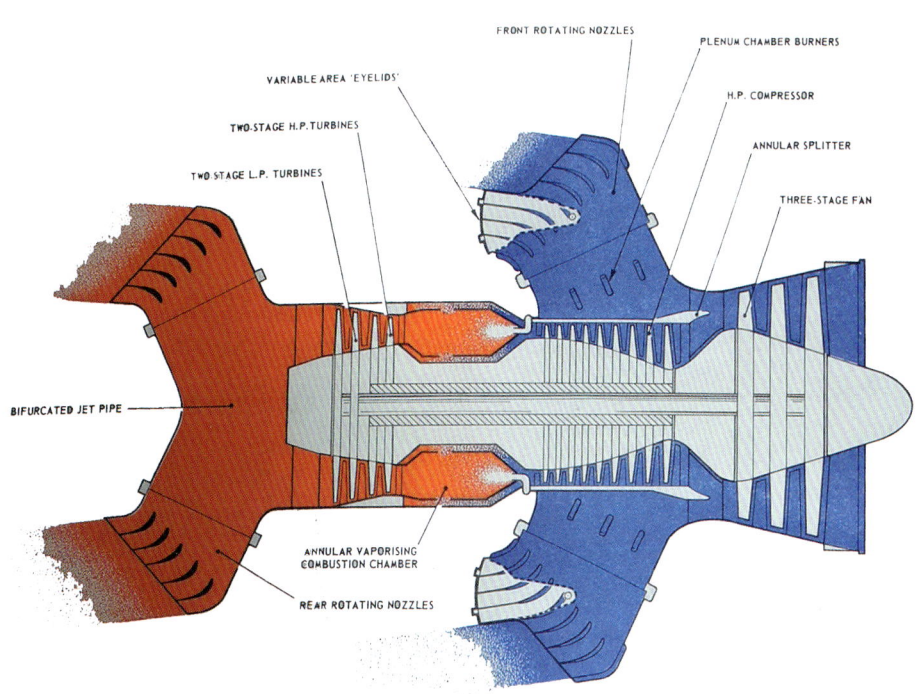

PLENUM CHAMBER BURNING OFF ('EYELIDS' CLOSED)

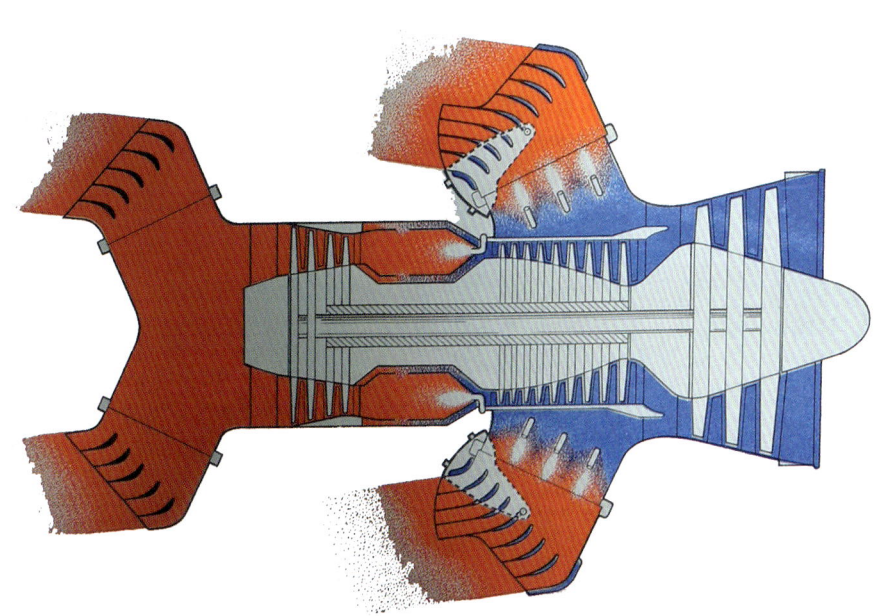

PLENUM CHAMBER BURNING ON ('EYELIDS' OPEN)

FIG.3.2. GAS FLOW DIAGRAM FOR BRISTOL SIDDELEY BS.100/9

BS.100/9 engine operation with PCB off and PCB on.

BS.100/9 engine showing the construction of the front starboard PCB nozzle.

Partial Pegasus engine test-bed undergoing PCB testing as part of BS.100 engine development.

P.1216 artwork showing the aircraft in RAF and Royal Navy colours.

Artwork showing Royal Navy P.1216 in STOVL mode clearing the ramp of a British carrier.

P.1216 artwork showing the aircraft in RAF and USMC colours shadowing Soviet bombers.

P.1216 artwork showing alternative stores fits on the boom pylons.

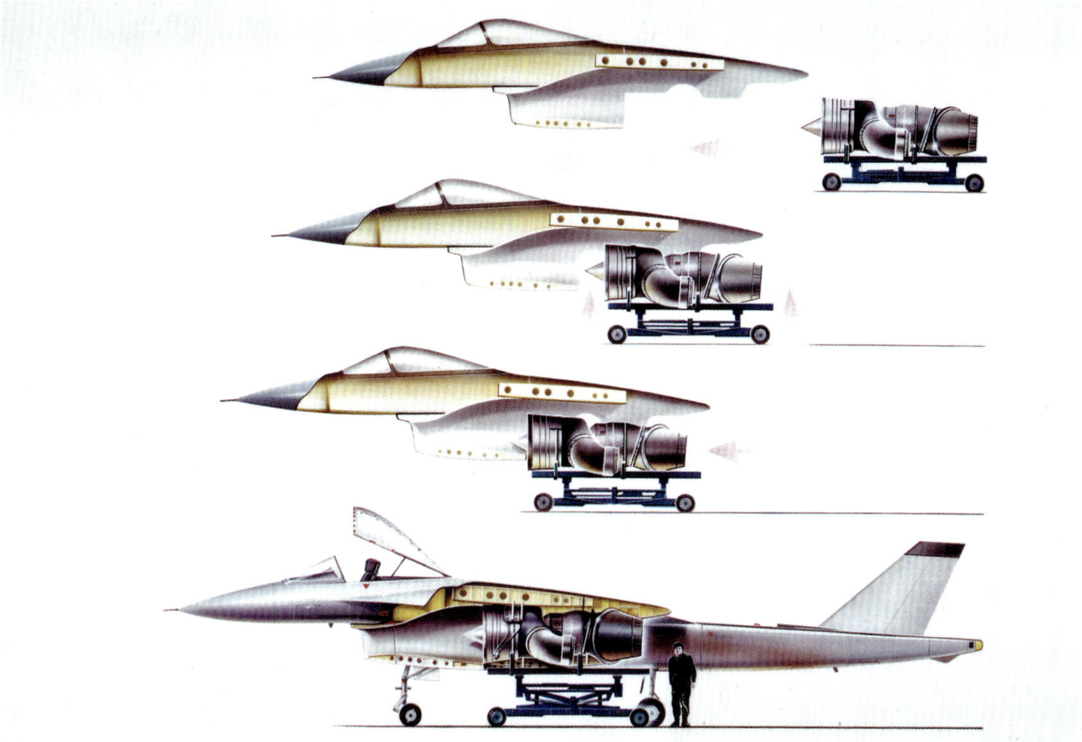

P.1216 engine change procedure emphasising the ease of service access.

P.1216 low speed wind tunnel model.

The Prime Minister, Right Hon Margaret Thatcher, inspecting the P.1216 mock-up during a visit to Kingston in 1982.

Going Up: V/STOL Studies 115

P.1154 (NAVAL)
TWO SEAT
SUPERSONIC INTERCEPTOR/STRIKE AIRCRAFT.
SHORT TAKE OFF / VERTICAL LANDING
B.S.E. BS 100/9 VECTORED THRUST TURBOFAN WITH P.C.B.
SPAN 30 FT. WING AREA 272 SQ. FT.
O/A LENGTH 57 FT. 9 INS WING L.E. SWEEP 42·5 DEG.
INTERNAL FUEL CAPACITY 1400 GALLONS

that the P.1154/1 gave the RN a reasonable interceptor but left the RAF with no V/STOL performance, whilst the P.1154/2 was a viable V/STOL aircraft as well as a naval interceptor of very useful performance.

In July 1963, in an attempt to justify the continued insistence on a common design for both services, the Minister of Defence had claimed in cabinet that doing so would achieve a saving of some £150 million in the next ten years. However, it was soon apparent that this figure involved significant sleight of hand, since it depended upon any RAF version not entering service until 1972/73, while the Air Staff expected a Hunter replacement to be entering service in 1969/70. It was pointed out that entry into service in the 1969/70 timescale achieved a saving of only £10 million. Such are the fudges by which politics is performed.[12]

Following further discussion with the MoA, the subsequent brochure, issued in August 1963, revealed an aircraft with simplified design now measuring 58 feet 6 inches with a span of 30 feet and a wing area of 287 square feet. Fuel capacity of 1,600 gallons was carried internally to power the BS.100/8 engine. The second crew member was deleted along with the high Mach shock-intakes

Above and below: P.1154 mock-up in Experimental Department at Hawker Siddeley Aviation, Kingston upon Thames, showing the drooped and trailed nozzle arrangement.

that the RN wanted, but with catapult facilities, wing folding and tricycle undercarriage retained which increased the weight for the RAF and reduced range and disposable load, eroding any usefulness to the RAF in the strike role. Clearly the entire project was heading for the rocks, resulting in a number of options being examined, including complete cancellation, alternative aircraft purchases and acceptance of two dissimilar P.1154 aircraft.[13]

At a meeting with the ministers of Defence and Aviation in October 1963, HSA representatives were informed that the bi-service design was unacceptable to both the RAF and RN, ideas to make the aircraft more acceptable being aired by the Kingston team.

On 31 December, the Chief of the Air Staff issued a brief appreciation 'Simplifying the P.1154' by the Air Ministry considering just what type of aircraft would suit the RAF as a minimum requirement. 'The primary purpose of the Hunter replacement will be to provide the air support that the Army requires … To do this it must be able to provide reconnaissance, and to attack and destroy a wide variety of targets threatening Army operations.' To achieve this as economically as possible, simplicity and ruggedness would be of great importance to enable the aircraft to operate in the field as far forward as possible. To this end, CAS proffered an aircraft requirement stripped of complex equipment.

The basic standard of the P.1154 under this concept would therefore be:

a) Airframe designed primarily for the low-level role i.e. simple air intakes.
b) Inertial platform plus simple navigational computer.
c) Small pulse radar, giving air-to-ground and air-to-air ranges and some information to the pilot to warn the pilot of high ground ahead.
d) Ability to carry two Red Top missiles.
e) A detachable pod of cameras for photo-recce.
f) A simple auto-pilot.

With such an aircraft in prospect, it was believed that recently suggested facilities could be dispensed with including:

a) Double shock air intakes to improve top speed at altitude.
b) Any form of AI radar.
c) Any requirement for auto-terrain following.
d) All-weather reconnaissance pod of radar and line-scan.
e) Possibly, the need to fit and clear the aircraft for carriage of nuclear weapons.

P.1154 mock-up in Experimental Department at Hawker Siddeley Aviation, Kingston upon Thames.

Although he did not know it, CAS was describing something very like the aircraft that the RAF would eventually receive six years later (i.e., the Harrier).[14]

As it became clear that the insistence on a common aircraft was driving up weight and complexity while simultaneously driving down possible radius of action and speed, and that commonality was now political doctrine incapable of reversal, particularly since Thorneycroft, the Minister of Defence, had announced in Parliament that both services HAD agreed a common design for the Hunter/Sea Vixen replacement, Ministry and Air Staff officials were left to grapple with the alternatives. Fear that the RAF would, as they saw it, have the P.1127 foisted on them was evident from September 1963; 'we shall be under pressure from the MoA to accept the P.1127 in some form in order to keep British industry in the business.' AVM Emson (ACAS OR) noted that 'if we both reject the P.1154/2, it will be a bitter blow to the Minister of Defence which he will not feel inclined to accept without the firmest professional advice and an alternative programme which will satisfy the difficult financial position.'[15]

P.1154 mock-up of twin-seat trainer version in Experimental Department at Hawker Siddeley Aviation, Kingston upon Thames.

As it happened, the Royal Navy had just such a proposal up its sleeve. An alternative purchase of the F-4 Phantom II was being mooted in the latter part of 1963, being seen as one alternative to the RN P.1154 buy. By the end of the year a joint statement by the Naval and Air Staffs to the Minister of Defence reported that the preferred solution was for P.1154 to be developed for the RAF and a purchase of F-4 Phantom to be agreed for the Royal Navy, fitted with Rolls-Royce Spey engines.

Hawker's final bi-service submission under P.1154 had offered an aircraft of length 57 feet 7 inches, span 28 feet 4 inches, wing area 269 square feet and sweep of 41.2 degrees, powered by the BS.100/8 with fuel capacity of 1,200 gallons. But it was a straw in the wind; the Royal Navy intention to purchase Phantom had doomed any bi-service design.[16]

As far as the Royal Navy's appreciation of the proposed P.1154 aircraft went, any work carried out by HSA to tailor it to the Royal Navy's requirements was superfluous, the Admiralty having decided that acquisition of the Phantom was their goal though this could not be confirmed publicly for political reasons.

Indeed, when a visit to the US by a party comprising Royal Navy personnel was arranged in late 1963, Peter Thorneycroft, the Minister of Defence, minuted the First Lord of the Admiralty to admonish him due to the real reason for the visit becoming widely known in Whitehall, much to the MoD's chagrin. 'I am anxious that this visit should attract the minimum of publicity and that if it should leak, it should be clearly seen to be a mission of technical enquiry related to the possible suitability of the Phantom for operation from British aircraft carriers. I sent you a minute yesterday in which I indicated the line … to be taken in the event of a leak.' Inevitably, in December, this proposal did leak to the press, much to the embarrassment of the MoD and service chiefs.[17]

As noted above, any proposed Phantom purchases would likely be powered by the Rolls-Royce Spey, off-setting to some extent the financial and political cost of such a purchase. The waters had been further muddied by a speculative proposal from Rolls-Royce to fit two Speys in place of the BS.100 engine in P.1154. Having spent the past two years briefing against the vectored-thrust engine in favour of their own dedicated lift engine and found no takers, Rolls-Royce were now becoming alarmed at the success of Bristol Siddeley, not least in the field of VTOL. Their Olympus engine was now powering the Vulcan fleet, was selected for the TSR.2 aircraft and, in developed form, would be selected for the Concorde programme. The Rolls-Royce proposal, particularly the costings

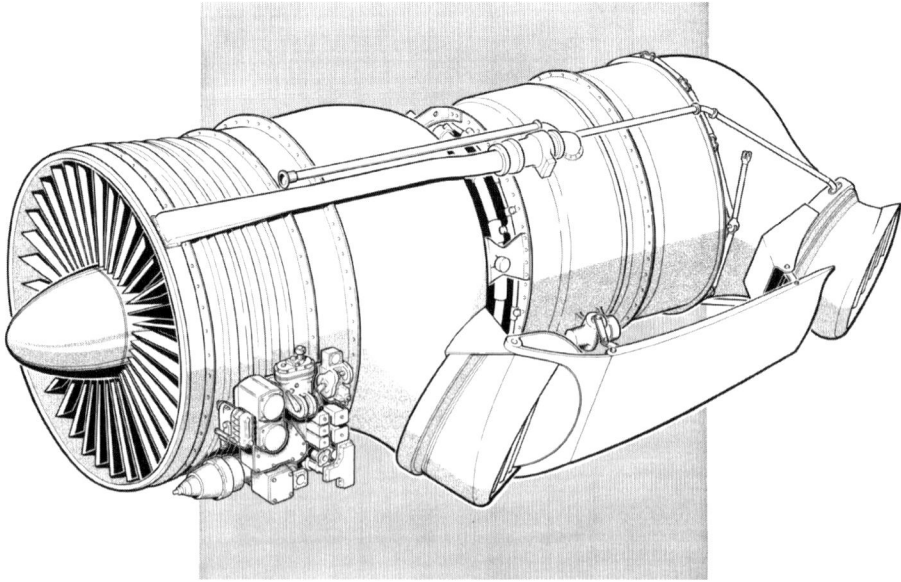

Bristol Siddeley BS.100/9 engine diagram showing the ramp arrangement for the control of front nozzle efflux.

submitted, was widely seen in Government as an attempt to 'buy in' to the programme at an unrealistic level. Although not welcomed at Hawker, the idea had adherents in influential positions and work was required to investigate it.

Work proceeded in the Project Office to scheme such a powerplant installation though there was little belief within HSA that such an arrangement would be chosen. As research continued at Rolls-Royce, the advantages of the twin-Spey arrangement lauded by Rolls-Royce and certain MoA officials dimmed, a requirement for 1,600° k reheat initially promised by the company being dropped, first to 1,400° k and then to 1,200° k and a need to increase the engine size leading, in the words of Sir Solly Zuckerman, to a situation where 'the gilt had now appeared rather to have gone off the gingerbread of the twin Speys'. On 26 February 1964 the matter was finally put to bed as far as Bristol Siddeley was concerned when Peter Thorneycroft announced in the House of Commons that any RAF P.1154 would be powered by the BS.100/8 engine.[18]

When it became obvious that HSA could not offer a design written around the Royal Navy specification, it felt free to reject P.1154 and turn to the aircraft it had wanted from the beginning – the McDonnell F-4 Phantom II. This was announced in February 1964, the American purchase being accepted with the sop that equipping with Rolls-Royce Spey engines would provide some UK offset.

With the Royal Navy now out of the picture, the RAF, and Hawker Siddeley, were free to revert to the aircraft that the RAF wanted, a single-seat supersonic V/STOL strike fighter. At last, the project could move ahead on a firmer footing and began to make great strides. It is worth highlighting that, given that the RAF P.1154's main role was to perform lay-down attacks behind enemy lines flying at high speed and low level to evade enemy defences, a supersonic capability was hard to justify, shades again of the virtual mania for supersonic performance whether actually required or not.

Be that as it may, to speed up the progress, the RAF showed significant flexibility in their requirements, for example dropping digital-computing and automatic-terrain-following in favour of an analogue nav-attack system and manual low-level navigation, removing the provision for Bullpup and Martel carriage etc. With the issue of a revised OR.456 in April, work began at Kingston and Hamble on production of jigs and detail parts, and plans for prototypes to undertake initial flight testing with first flight in 1966.

The development programme envisaged eight development aircraft with first flight two years from Instruction To Proceed (ITP), that is, by July 1966 at a cost of £170-200 million, approval being given in November 1964 to proceed. However, first flight of a fully representative aircraft fitted with radar and navigation system looked to be increasingly delayed due to a spat between Ferranti and the Ministry of Aviation over perceived excess profits due from Ferranti on the Bristol Bloodhound missile programme, causing concern at Hawker Siddeley. It was expected that the RAF would order 157 single-seat aircraft and a further 25 twin-seat trainer variants at £1 million and £1.2 million respectively.

With initial testing of the BS.100 engine completed in September 1964, in October, the various changes to the design were formalised in specification SR.250D for an aircraft of 57 feet 6 inches and span of 28 feet 4 inches with wing area of 269 square feet swept at 41.2 degrees on the leading edge. Empty weight was calculated at 20,100lb and maximum all up weight at 40,050lb. The definitive engine would be the BS.100/8 Phase 2 able to be replaced from below and offering 33,900lb thrust with PCB (or 35,900lb for twenty seconds using the short lift rating). Maximum thrust without PCB was 26,200lb. The variable geometry air intakes were supplemented by blow-in doors while the exhaust flow from the front PCB nozzles was controlled by adjustable ramps between the front and rear nozzles. A maximum speed of Mach 1.7 was forecast at altitude, the aircraft being equipped for either strike at low level using inertial navigation driving a moving map display coupled with a head-up display and carrying a variety of conventional ordnance, or interception at altitude with Red Top infra-red guided missiles.

As work continued at Kingston on production of detail parts and the first airframe modules in the jigs, manufacture began at Hamble of the first wing sets. All now looked set for the first aircraft to be available in late 1966. Meanwhile, in the fragile world of politics, ominous changes were afoot, the Conservative administration of Sir Alec Douglas-Home being ousted from office on 16 October 1964 by Harold Wilson's Labour Party arriving intent on making largescale economies, particularly in defence, to support its various social programmes. Within the RAF and the Air Ministry the writing appeared to be on the wall for at least one major defence project, and urgent assessment was undertaken to ensure that the TSR-2 programme should survive whatever calamity might be heading their way. If necessary, P.1154 could be sacrificed if it saved TSR-2.

As a harbinger of things to come, on 8 January 1965, Roy Jenkins, the new Minister of Aviation visited HSA Dunsfold and Kingston with a party including AVM Hughes to receive a briefing on P.1154 and Kestrel. It was, he declared,

Going Up: V/STOL Studies 123

Both images above: Bristol Siddeley BS.100 engine under construction at Patchway, Bristol.

his avowed intent to 'buy in the cheapest market regardless of the requirements of home industry'.[19]

Further uncertainty regarding the UK's future defence policy was generated on 15 January by a dinner at Chequers, given by the new prime minister, to which the heads of the country's aircraft industry, including HSA Chief Executive Sir Arnold Hall, were invited, Camm travelling up to London to hear at first-hand what had been discussed. The following week, Hall again visited Harold Wilson in an effort to obtain some idea of just what the Government's intentions were with regard to HSA aircraft (i.e., P.1154 and AW.681). The answer was considered 'reasonably satisfactory', not quite the unambiguous response that Hall would have been looking for and an assurance which would not last a month. For the first time, a suggestion that the RAF would take P.1127s and, by inference, lose P.1154 was floated by AVM Hartley to Hawker Siddeley representatives but they would insist on Phantoms as well. The next day Sir Arnold Hall, with Mr Lidbury, met with Roy Jenkins. Clearly so much interaction with politicians was not good news.[20]

On 2 February 1965, Denis Healey, as Secretary of State for Defence, announced the cancellation of P.1154 and, for good measure, the AW.681 VTOL transport project. The RAF would get some Rolls-Royce Spey powered F-4 Phantoms and a small batch of developed P.1127s. TSR-2 was saved – perhaps. In the event, this programme was also cancelled once the first aircraft began its flight development phase, the cancellation of the three major aircraft projects leaving the UK aircraft industry bereft of any significant work programmes and leading to widespread redundancies in the industry. Sir Arnold Hall reported that up to 14,000 jobs were likely to be lost as a consequence of the government's swingeing cuts, George Brown blustering in Parliament that this couldn't possibly be justified. The irony of all this was, of course, that many who did lose their jobs were the very people who had voted Wilson's government in. At Kingston, an assessment of loading in the Design organisation revealed that 100 staff would be made redundant immediately, while the Hamble Design Office would close completely, representing a loss of approximately 20 per cent.[21]

The claim that P.1154 was being cancelled due to excessively high costs would sit ill with Hawker Siddeley, and with good reason. In November 1963, when this had first been raised, the company had been puzzled since it had strenuously worked to keep costs down. Whilst at first promised by the Minister of Aviation that they could see the costings, subsequent discussions with MoA

P.1154 wings under construction at Hamble.

officials failed to produce detailed figures, leading to the suspicion that MoA costings were not all they claimed to be. A Hansard report dated July 1974 set the whole question of aircraft development costings in a clearer light. It showed that the P.1154 project had cost the country £21 million and HS.861 £4 million, compared to TSR.2, which had cost the country a whopping £178 million while its successor, the F-111, cost £13.5 million, much of it in cancellation penalties. The government had therefore spent a colossal £216 million on programmes which had produced precisely no aircraft.

So it was that the V/STOL 'ball' passed back from Fozard to Hooper who now had the urgent task of preparing the development P.1127/Kestrel project for service with the RAF, though with Fozard now jobless again he was appointed Chief Designer P.1127 (RAF) in September 1965.

P.1154 was a radically different way of addressing the requirement for an interceptor/strike aircraft capable of dispersed operation, so why did it fail? Perhaps a better question might be, could it have succeeded? The answer to that is yes, success would have been achieved. The knowledge and technology were to hand and, in another age, would have seen the aircraft enter operational service. But it was conceived in an era of austerity, an era which saw the UK's place in the world changing and shrinking.

The most innovative element of the design was its engine. Along with the later Harrier and its progenitors, it was an aircraft literally designed around its engine. If the engine development failed, then the aircraft programme failed too. Could Bristol Siddeley have brought the PCB-powered BS.100 to maturity? Again yes, though progress would probably have been slower than Stanley Hooker and his colleagues hoped. Cancellation of P.1154 of course meant cancellation of the BS.100 engine as well. Dow noted that, after the event, at Bristol 'the demise of BS.100 was met with a quiet sigh of relief. There were many who were all too well aware that PCB had not been run on a BS.100, nor flown, though PCB rig testing HAD been conducted on a Pegasus 2. The effectiveness of the ramps to reduce the size of the front nozzles had yet to be proved, and the ground erosion and hot gas reingestion problems expected with PCB had not been evaluated in practice.' PCB had been trialled at Aston Down using Pegasus engine 906 in November 1964 and had shown good progress. Of course, all these issues were dependent on funding being available, which, in the final analysis, it was not.[22]

Of the three pertinent points highlighted above, with the benefit of hindsight and extensive Harrier operations, what would have been the outcome? The ramps positioned behind the front nozzles were designed to impinge on the nozzle efflux in supersonic flight to match engine output to flight conditions, carrying out the same function as the more standard translating nozzle on 'normal' reheated engines and should not have proved too onerous to bring up to production standard. The question of ground erosion was more problematic. The later Harrier possessed a relatively benign exhaust flow from the front 'cold' nozzles and a rather hotter exhaust from the rear nozzles. Extensive operation allowed a body of knowledge to be built up, allowing erosion issues to be minimised in service. However, the exhaust from BS.100 was of an altogether different magnitude. The aircraft's centre of thrust was completely different to Harrier due to the greatly increased thrust from the front 'reheat' nozzles. Since the centre of thrust was designed around PCB generated thrust, reheat had to be used for VTOL and landing manoeuvres, giving rise to greatly increased heating and erosion problems which may have proved difficult to overcome in dispersed operations. Off-site operations of this type would certainly have been markedly different to those of the later Harrier.

Hot gas re-ingestion would also have proved troublesome, given the greatly increased energy involved in hovering and vertical landing manoeuvres with the BS.100. However, again these would have been overcome through trials and a worthwhile performance regime produced. Ultimately, these concerns would be addressed in far more detail as part of the later efforts to design an ASTOVL fighter in the 1980s.

The Royal Navy was never enthusiastic about the aircraft. It simply did not fit their requirement for a supersonic interceptor with good loiter performance capable of launch and recovery from its existing carriers. While the new breed of 'super-carriers' epitomised by the long awaited CVA-01 could certainly have been designed to optimise use of V/STOL aircraft, there was the potential danger of the service losing out on others of the class since V/STOL meant no need for large floating airfields, with their concomitant large crew complement and imposing spectacle, not something that the Sea Lords could contemplate. In the event, although the Royal Navy won the battle in obtaining the Phantom II, they lost the war in failing to obtain the new carriers, CVA-01 being scrapped in 1966, and for a while being threatened with having no fixed-wing air complement at all, until reluctantly accepting the Sea Harrier and small 'through-deck cruisers' of the Invincible class.

Ultimately, the failure of P.1154 can be traced to the initial insistence of the Treasury that the RAF and Royal Navy share a common aircraft to fulfil their completely different requirements. The idea of a joint aircraft was to save money that in the late 1950s/early 1960s was simply not available. Cost saving governed almost all political thinking on the subject, it being admitted even by Peter Thorneycroft in October 1963 that it might be necessary to cancel one of the three big aircraft projects – TSR.2, AW681 and P.1154 – since the defence budget could simply not afford the development costs of all three. While it was Harold Wilson's administration that bit the bullet and initiated cancellation within the defence budget, the loss of *all three* projects had never been contemplated by the Conservative government. It would have been politically unthinkable.

Hawker Siddeley was perfectly capable of designing a new aircraft to satisfy each service but, ultimately, was unable to offer one aircraft that answered the disparate needs of both parties without compromise on their part. Indeed, it is unlikely that any other company could have done so. Hooper's assessment of P.1154, written in 1974 concluded, 'Whether the P.1154 contained too large an element of run-before-walk we shall never know. But inter-Service rivalry and Ministerial dreams of commonality of hardware in spite of harshly divergent operational requirements, left it exposed to die by the political axe. We lost perhaps three years on the present state-of-the-art with the Harrier during this period.'

As a postscript to this saga, one may turn to Ralph Hooper's mature conclusion following his retirement. In considering alternative layouts to the P.1216, dealt with in a later chapter, he thought that 'The "1154" layout, I now see as dishonest as it is running headlong into a problem which we have seen worsening … It is not just the temperatures … and the acoustics – those can be lived with. But why build a modern long life jet aircraft that persists in shaking itself in imitation of its reciprocating forebears?'

Chapter 6

SABA: Small Agile Battlefield Aircraft

Close Air Support (CAS), the use of offensive aircraft to support troops in tactical movements on the ground, has traditionally been the poor relation of aircraft procurement and seldom have aircraft been ordered specifically for this role. More often fighters have had this duty added to their roles as each generation rediscovers the requirement in time of conflict.

The 1960s saw the role of helicopters developing from lightweight transports to include an offensive close-air support role, particularly in the Central European Theatre, both NATO and the Warsaw Pact producing designs to support this role. Within NATO this effort had mainly been carried out under US Army control, which required a weapon able to combat the massed tank formations expected to be a major component of any attack from the east. The proliferation of the US Army anti-tank helicopter proved to be a concern for the US Air Force, which saw its dominant nuclear-armed stance in Europe threatened by its ground-based competitor service. In an effort to recapture what it saw as its traditional role, a USAF competition eventually led to a design for a relatively cheap, heavily-armed and armoured close air support aircraft, the Fairchild Republic A-10 Thunderbolt II, which came into service from 1976. Powered by two turbofans, the straight wing, subsonic design was intended to be cheap to build and operate and capable of absorbing significant damage while carrying out its task. Main armament was a 30mm rotary cannon in the nose and a variety of missiles and bombs carried on wing pylons. This successful aircraft was scheduled for replacement several times throughout its career but continued in service, finding ample service windows in the Gulf War of 1990 and in Afghanistan and remains in service to date.

Post-war thoughts on a close air support role within the UK had previously centred around the ground-attack versions of the Hunter, but this aircraft was not optimised for modern combat threats of the type that were developing in Europe in the 1970s, particularly that posed by the offensive helicopter and thoughts on a modern replacement emerged at times from the Hawker Project Office. Mainly, such schemes were modifications to what had originated as fighters;

during the war this had meant various Hurricane marks and the Typhoon. In the post-war era the Hunter was developed as a ground-attack aircraft – the Hunter FGA.9 – which successfully plied its trade for many years. However, the Harrier was to be employed from the outset as an effective ground-attack aircraft rather than fighter; its short reaction time, thanks to the forward basing of the aircraft, making it a potent ground-attack/close air support element within NATO forces in the European theatre. While the Harrier was an effective ground-attack aircraft, it was not cheap. What was required was an effective *and* cheap aircraft capable of surviving the bruising environment that close air support represented.

To this end, in August 1973, ideas for such an aircraft had polarised around project code HS.1196 under the term 'Minicas' (minimum size/cost, close air support) which, in a presentation to RAE Farnborough officers in March 1974 by John Allen, Chief Future Projects Engineer, Kingston, was described as 'a battlefield attack aircraft to replace Jaguar and Harrier'. Having failed to find a suitable layout using conventional ideas, a new approach was initiated aimed at producing 'a low-cost aircraft while also improving the kill chance against a tank in a first pass attack'. To achieve this, steps were initiated which included:

a) A specialised aircraft designed solely to kill tanks in daytime.
b) Reduced size by use of GW rockets.
c) Use of fuselage bomb bay to reduce drag and simplify wing and fuselage construction.
d) Enhanced range, short field and better performance with Rolls-Royce RB.401 engine.
e) Engine mounted above fuselage to reduce size and reduce installation costs.

It was hoped that such a basic aircraft could be produced for approximately £300,000 at 1974 prices. Minicas would have been of a size somewhere between the Gnat and the Hawk had it proceeded which it did not, the project joining many others aimed at low cost/high capability. Thereafter, little was heard of the CAS concept within the company and it was not until the 1980s that thoughts on a dedicated close air support design reached a more solid foundation.

On 14 July 1983 Ian Hall, Director of Engineering and Project Assessment at Kingston, wrote to Syd Gillibrand, Regional Managing Director, and Len Milsom, Divisional Director of Projects within BAe, regarding a meeting of the Marketing and Business Development Committee which had considered possible roles for the company's Hawk 200 light strike aircraft, then in prospect. Hall wished to suggest using the aircraft in an anti-helicopter role. Given the increasing preponderance of attack helicopters such as the Hind within the Soviet inventory and their use in attack 'swarms', Hall felt that the single-seat Hawk could prove to be a useful counter to this type of threat. Clearly a certain degree of crystal ball gazing would be required to answer this query since the Hawk 200 prototype would not fly for another three years, but the idea certainly proved of sufficient interest within the company to begin debate on a counter-helicopter aircraft based on the Hawk. The following week Johnnie Johnson in the marketing team confirmed that this was a role being considered for the Hawk 200, the plan to fit radar and twin cannon being seen as particularly suited to an anti-helicopter role.[1]

Notwithstanding lukewarm interest within the Air Staff for a similar concept presented by French industry based on Alpha Jet, Ralph Hooper had alerted ACAS/OPs to the subject and received a more positive response. At this point, Chris Farara, Hawk Project Manager, in confirming 'high level interest' in the idea, asked that the concept be formulated in a concise paper such that a development programme might be assessed and costed.

P.1238

The concept of helicopter suppression as a definitive role then moved away from the Hawk towards a completely new airframe. This was, in large part, due to outline work carried out by Andy Jones, then Chief Test Pilot at Dunsfold, who foresaw a possible entry for this concept in the MoD's AST.412, released to industry in the search for a new basic trainer to replace the Jet Provost. Jones suggested in August 1984 that if a low-cost attack aircraft could be built that allowed use as a trainer in a secondary role, with strike as the primary role, then the company might avoid producing yet another trainer in an already crowded market and offer a budget operational attack aircraft into the export market.

The appropriate role might be, he suggested, helicopter suppression, this being an area with an increasing threat but little counter. Given the very different flight characteristics and high manoeuvrability of the helicopter, particularly at low altitude and speed, Jones saw the main requirements for a 'helicopter killer' as 'highly agile turboprop powered … capable of achieving speeds of 300-350 knots, and turning at two to three times the rate achieved by fighter aircraft. Using an engine in the 1,500shp class it would have a combat weight

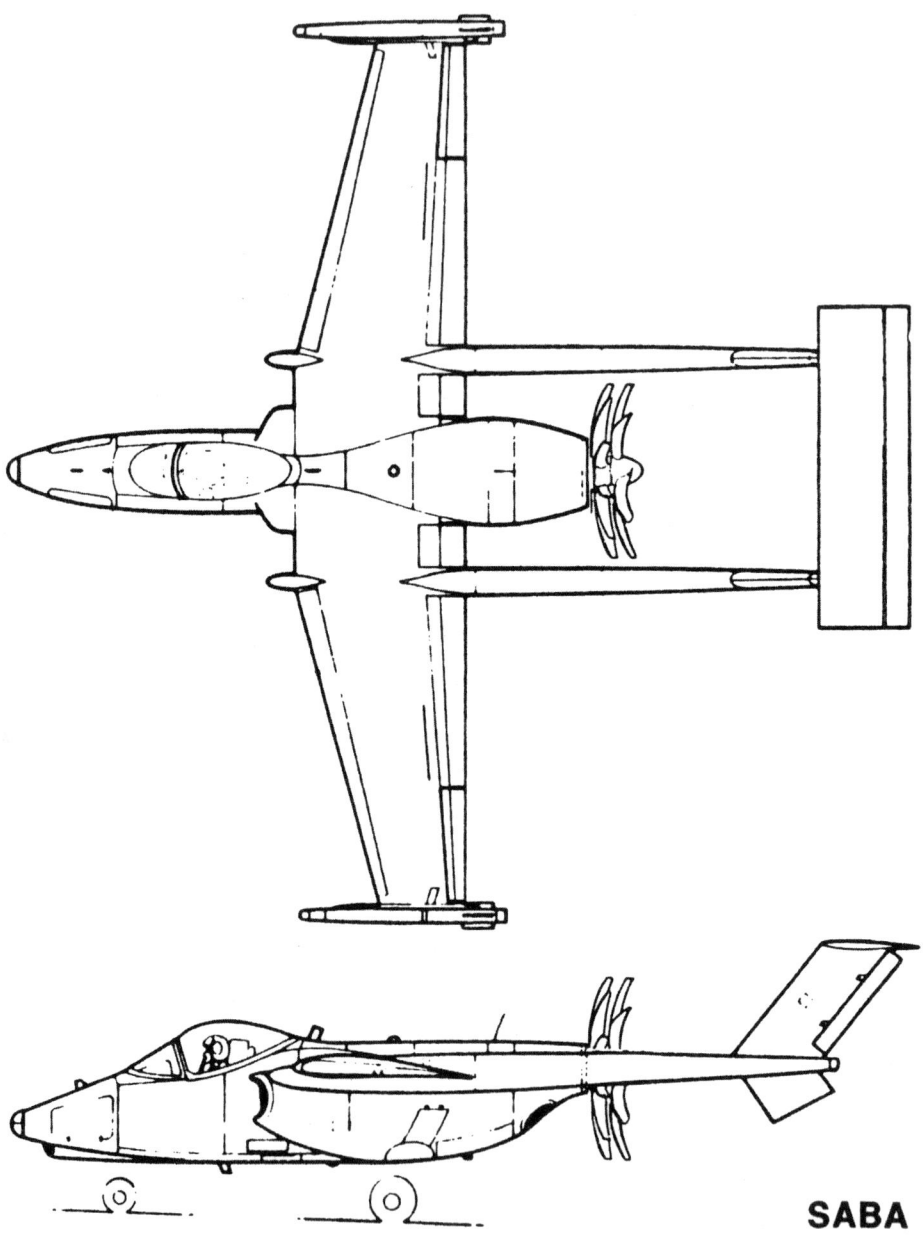

P.1238 SABA GA, powered by Avco Lycoming T-55 unducted fan.

of some 4 tonnes. Carrying a crew of two, equipped with advanced avionics and armed with four anti-helicopter missiles, the aircraft would be designed to have a maximum continuous cruise of 300 knots. As such it would pose a substantial threat to helicopters.' In the trainer role, Jones envisaged a reduced weight due to

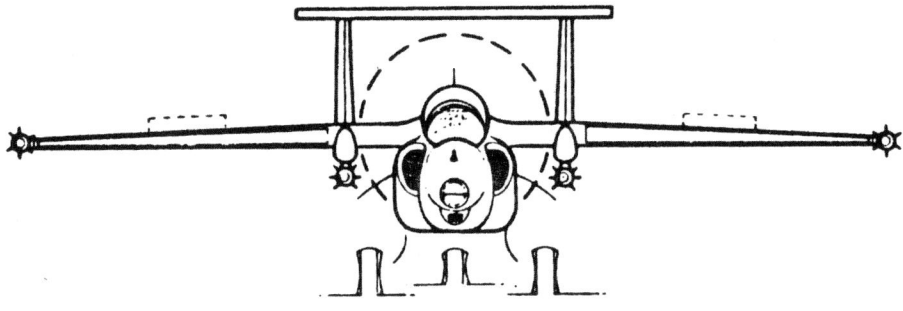

P1238

Length	11.55m
Span	10.97m
Wing Area	18.58m²
Aspect Ratio	6.5
MTOW	5034kg
	Avco Lycoming T-55
	3357 kW

removal of armament, simplified avionics and a smaller SHP engine giving an airframe of around 2.5-3 tonnes which should satisfy AST.412.[2]

Jones depicted an airframe layout that was 'relatively unconventional', featuring a twin-boom layout with a single turboprop driving a pusher propeller. It had a conventional low-mounted straight wing of 30 feet span and area of 160 square feet, twin fins, a length of 38 feet, a 4,000shp turboprop with internal fuel of 500kg allowing a loiter time of some 2.5 hours. The aircraft would be capable of an instantaneous turn of 35 degrees per second and a sustained turn of 30 degrees per second for five seconds.

Jones initiated discussions with Ralph Hooper, by then Deputy Divisional Technical Director, Weybridge Division and Harry Fraser-Mitchell, Assistant Chief Airframe Engineer, concerning just what design aspects were required for such a role. 'In the end it was decided that we wanted very high rates of turn combined with good acceleration/deceleration, dirt field capability and the ability to acquire and fire in a very short time. In turn this demanded what I would term a "High Tempo Cockpit" incorporating a lot of high tech assistance for the pilot.'[3]

In July 1985, Fraser-Mitchell reported back to Executive Director (Technical) Gordon Hudson who had requested a feasibility check of Jones' proposal. Fraser-Mitchell worked on the basis of 'the need for an instantaneous turn rate of 36 deg/sec plus the ability to transit at speeds up to about 450 knots at low level. Subsidiary but necessary requirements are for good control for "nap of the earth" flying, the ability to operate from short, semi- or unprepared airstrips and good visibility both from the cockpit and for target seeking sensors'. He further suggested that, to achieve the speed requirement, a multi-bladed prop-fan might offer a good solution. In conclusion Fraser-Mitchell stated that 'a design to meet the requirements formulated in the first section is indeed feasible'. This initial stab at a useful layout attracted the project code P.1238, out of sequence due to being numbered later in the design process. On 1 April 1986 Andy Jones wrote to Chris Hansford in the Future Projects Office to outline his thinking behind the SABA concept such that Future Projects could study the idea and scheme a realistic proposal around these thoughts. In reiterating the salient points – agility, high speed, rough field performance, easy handling, low cost – Jones admitted that 'All of the above is of course easy to say and perhaps less easy to do…The big question is whether these aims can be met to any degree'.[4]

P.1233

Work now began in Future Projects under code P.1233 and in May 1986 Vic Hancock produced the first Future Projects drawing of a potential SABA aircraft. This again depicted an unconventional layout: a straight wing spanning 10.97 metres with swept roots, set towards the rear of the 9.54 metre fuselage; large canard foreplanes adjacent to the cockpit, a contra-prop Avco Lycoming T-55 unducted fan mounted at the rear, together with dorsal and ventral fins and nose-mounted rudder; this layout, now officially coded P.1233/1, became the baseline for future studies. In amplifying the choices made for this layout it was stated that the configuration was chosen to 'maximise fuselage volume while retaining a pusher prop-fan. The aft prop-fan provides directional stability, at the cost of reducing the effectiveness of a rear-mounted rudder [hence the nose-mounted location]. The canard layout enhances manoeuvrability since the foreplane load required to increase angle of attack acts in the same direction as wing lift … The pilot is protected by cockpit armour and a low-glint canopy embodying flat panels of armoured glazing.' Another positive attribute of the pusher prop configuration was that it left the nose clear, providing excellent visibility for both pilot and sensors. However, Andy Jones recalled that 'I do remember a debate about the very high noise signature that would probably be inherent in the variant pushed by a 4,500hp motor driving contra-rotating props!' Not the stealthiest of designs perhaps.[5]

P.1234

While P.1233 studies would become the preferred avenue of research for a SABA product, studies were also carried out under P.1234 by Keith Towell and Vic Hancock in the Weybridge Future Projects office. The first of these, P.1234/1, examined what might be termed a 'reverse delta' blended wing configuration of 10.97 metres and an area of 35.95 square metres, carrying a cannon, aimed by helmet-mounted sight in a ventral turret blended into the 9.6-metre fuselage. Power was provided by a Rolls-Royce Adour turbo-fan using wing-root intakes. In describing this to Andy Jones, the pair noted 'We have retained the forward fuselage of P.1233, less the fixed gun, but have gone to a slightly unusual wing layout to see how it works'. P.1234/2 offered a more conventional, straight wing of 10.97 metres coupled to a stubby fuselage of just 9.05 metres carrying twin dorsal and ventral fins and tailplanes at the rear, the whole powered by an Avco Lycoming ALF-502 engine fed by wing-root intakes that dominated the design. P.1234/3 offered a fairly conventional tailless delta wing of 9.14 metres span and area of 30.75 square metres, powered by a Rolls-Royce Adour RT-172-871. However, its armament would be decidedly unconventional, being hyper-velocity missiles fired from a ventrally-mounted twin-tube rotary launcher.[6]

In December 1986 an internal report on New Business noted three possible concepts that 'merited detailed appraisal in the near term: a) an unmanned fighter aircraft; b) a maritime patrol aircraft; c) a small agile battlefield aircraft (SABA)'. Of the latter, the report noted that 'heavily armed helicopters are now being deployed in large numbers and the tilt-rotor will enter the field shortly. No convincing counter to either of these concepts exists, particularly in nap of the earth operations … It appears feasible to produce a low cost small agile battlefield aircraft able to counter these threats …

'Preliminary studies indicate that a canard aircraft with a combat weight of around five tonnes, propelled by an unducted fan, could offer genuine dispersed operation, a maximum speed of over 400 knots and unparalleled agility. Able to turn through 180 degrees in under five seconds, this aircraft would have a minimum radius of turn of some 70 metres, with a peak rate of turn of 40 deg/

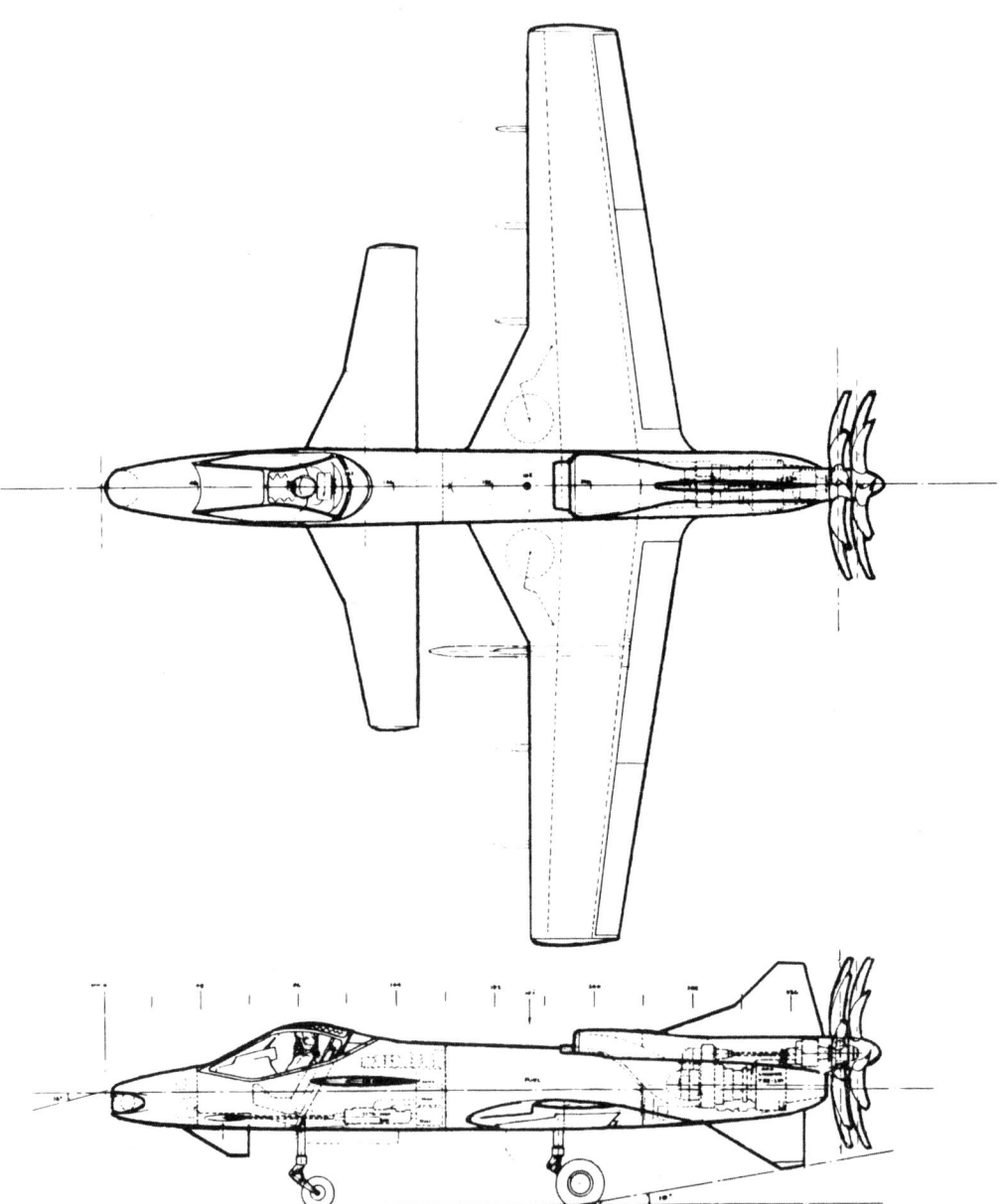

SABA

P.1233-1 SABA GA with large canard and nose rudder.

sec while carrying 6–8 air-to-air missiles. It is recommended that this concept should be studied in greater detail.'[7]

Further investigation of the SABA concept using the P.1233 baseline continued through 1986 with Jones, now Hawk 200 Sales Manager, having

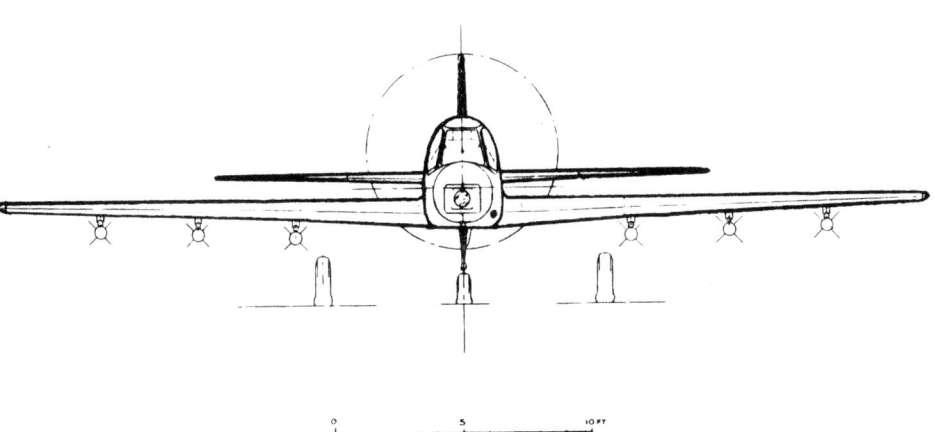

P1233 - 1

Length	9.54m
Span	10.97m
Wing Area	20.39m²
Aspect Ratio	5.9
MTOW	4989kg
	Avco Lycoming T-55
	3357 kW

retired from the Chief Test Pilot position, seeking assistance for assessment of the likely market for SABA from the Kingston sales and marketing team. P.1233/3 resulted in a more conventional wing planform with slightly swept leading edge. Engine air intakes were placed either side of the upper fuselage above the wing, the fuselage becoming somewhat 'chunkier' and stubby.

By August 1987 Keith Towell at Weybridge had drafted a full specification for SABA. This covered all configurations already considered with a view to selecting an optimum design later. The primary mission would be anti-helicopter with secondary mission of close air support (CAS). The aircraft was to be capable of sea-level speed of 400 knots with primary weapon load and a 180-degree turn in five seconds at full combat weight, approximating to an instantaneous turn rate of around 40 degrees per second, the turn radius to be within 125 metres. Loiter performance with full weapon load was to be approximately four hours, take-off to be achieved within 250 metres and a landing roll of 200 metres on an undercarriage capable of rough field performance.

Six baseline configurations were suggested: P.1233/1, P.1234/1, P.1234/2, P.1234/3, P.1238 and P.1239, the powerplant to be either a turbofan, such as a developed Adour rated at around 7,500lb thrust, a development of the Avco

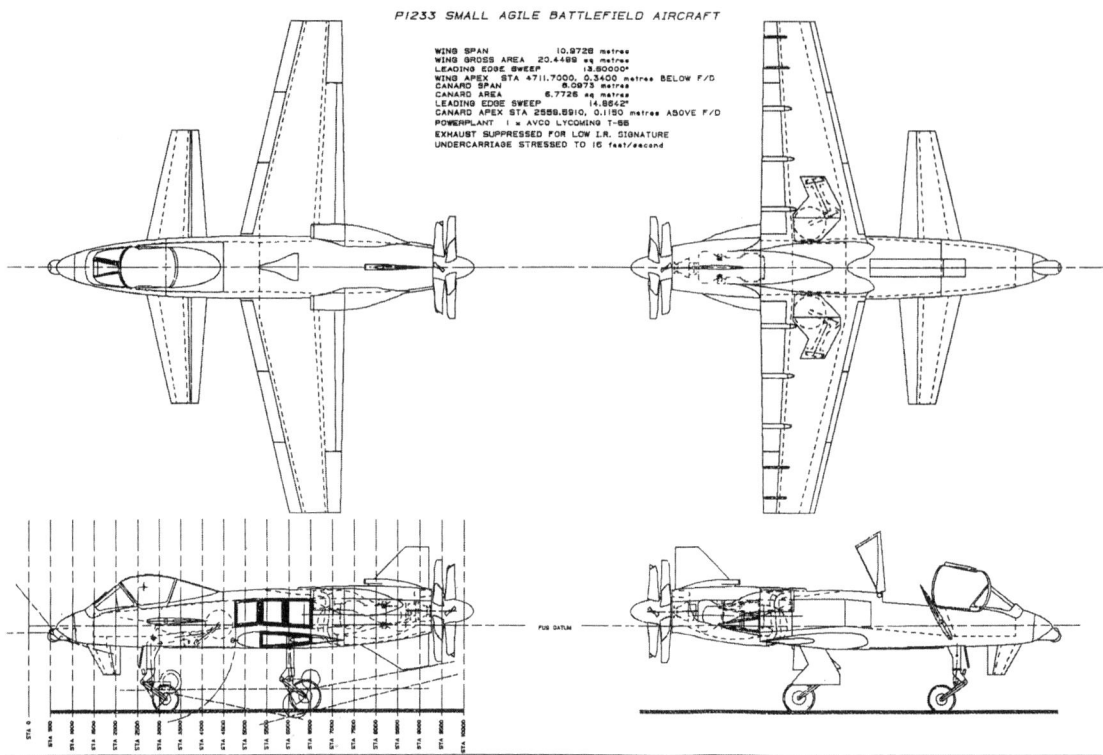

P.1233 GA with alternative wing planform.

Lycoming ALF502 or a General Electric TF-34. Alternatively, a turboshaft engine, such as the Avco Lycoming T-55, Turbomeca/Rolls-Royce RTM322, Allison T56 or General Electric T64, was suggested. Primary armament would consist of a fixed fuselage-mounted gun such as a 25mm cannon with 150 rounds of ammunition or 12.7mm MG with 150 rounds. Missiles would likely be six underwing ASRAAM or, alternatively, hyper-velocity missiles fired from a turret.[8]

As interest in the project gathered pace within the company, a series of review meetings began in October 1987, personnel being drawn from Kingston, Brough and Warton, and at the November meeting it was agreed that *Jane's Defence Weekly* for 5 December would carry a press release publicising the concept. Jones recalled, 'At one point it was thought that some publicity for the project might be a good idea and so I got clearance from two senior folk to let a bit of information out to the aviation press. The next thing I knew I was summoned to the office of Ray Lygo, who was Chief Executive, together with John Farley, to explain myself! I am not sure why John was invited – nor was he – but it was clear that

SABA: Small Agile Battlefield Aircraft

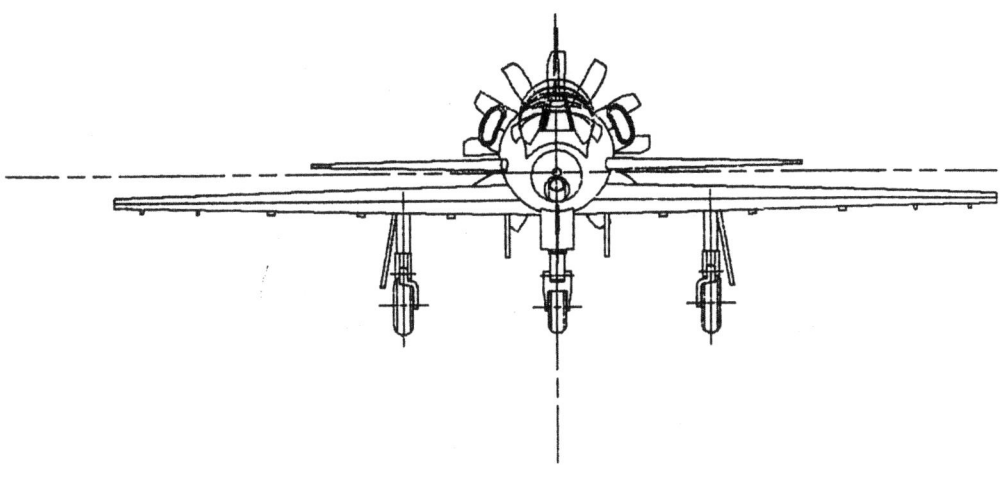

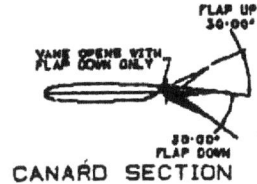

CANARD SECTION

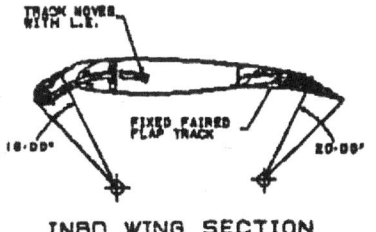

INBD WING SECTION

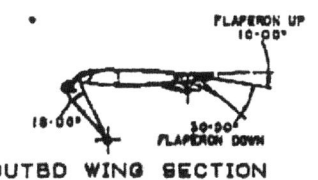

OUTBD WING SECTION

TITLE: PRELIMINARY GEOMETRY — SABA006

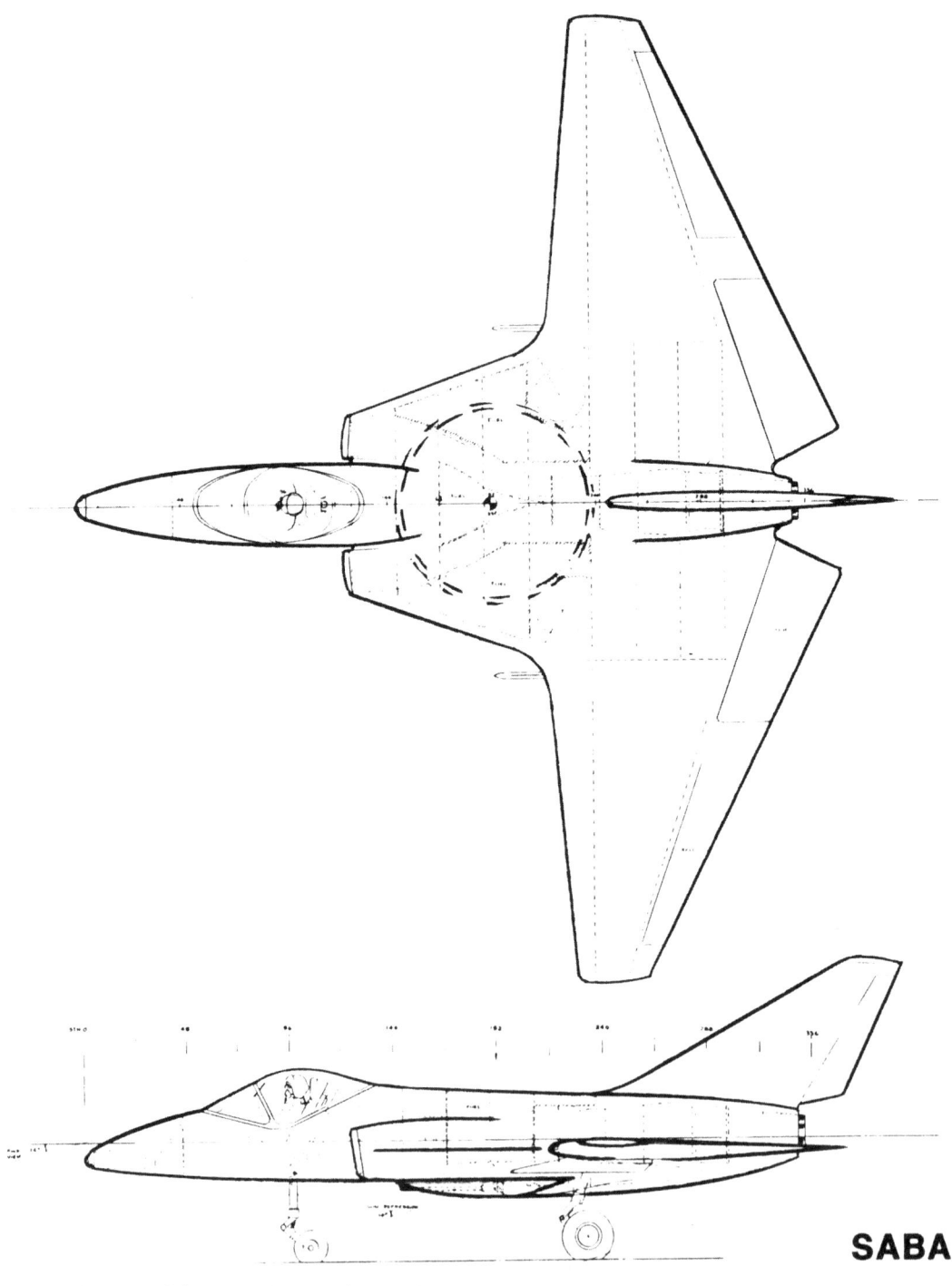

P.1234-1 GA, powered by RR Adour RT-172 turbofan and mounting a ventral turret.

P1234 - 1

Length	9.60m
Span	10.97m
Wing Area	35.95m²
Aspect Ratio	3.3
MTOW	5754kg

RR Adour RT-172-871
25.4 kN static thrust

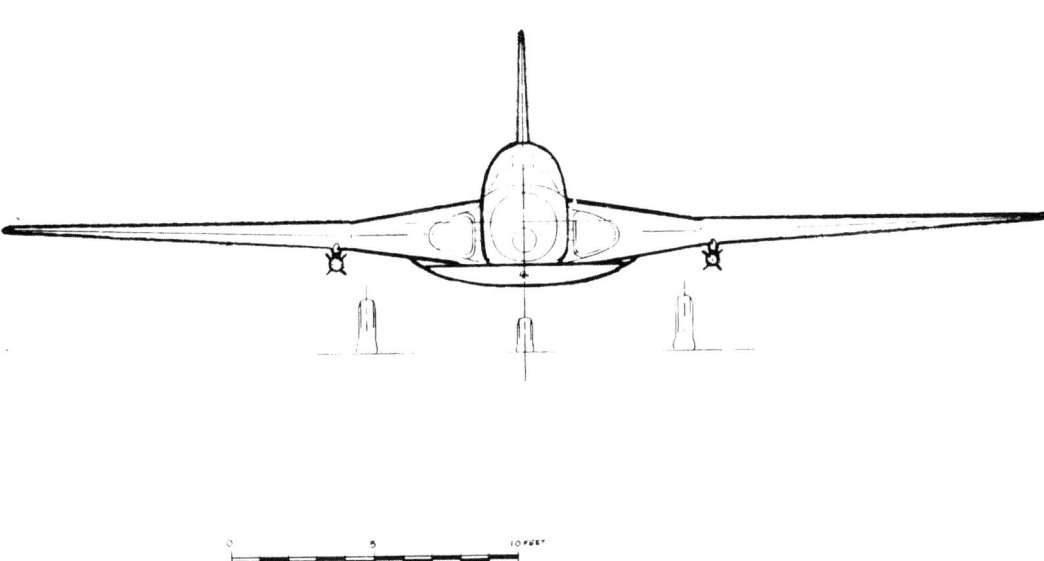

our chief was very angry. He was convinced that the project would damage prospects for the Hawk and concerned that the RAF had apparently had their noses put out of joint because they had heard nothing about the idea. He had evidently spoken with the individuals who had cleared the conversation with the press and they were said to have denied having done so. We were told to give a presentation to the RAF Operational Requirements people; we got a rather antagonistic hearing. No more was said and nothing more was heard either from the RAF or from Ray Lygo. But Lockheed very quickly produced artists impressions of a rather similar concept.'[9]

Preliminary discussions had begun with the US in the person of Deputy Under Secretary of Defense for Tactical Warfare Programs, D.N. Fredericksen, who had shown 'considerable interest'. Consideration was given to producing a demonstrator aircraft for 1992 though this would necessitate an early choice of aircraft configuration, assisted to an extent by wind-tunnel testing due to start

SABA

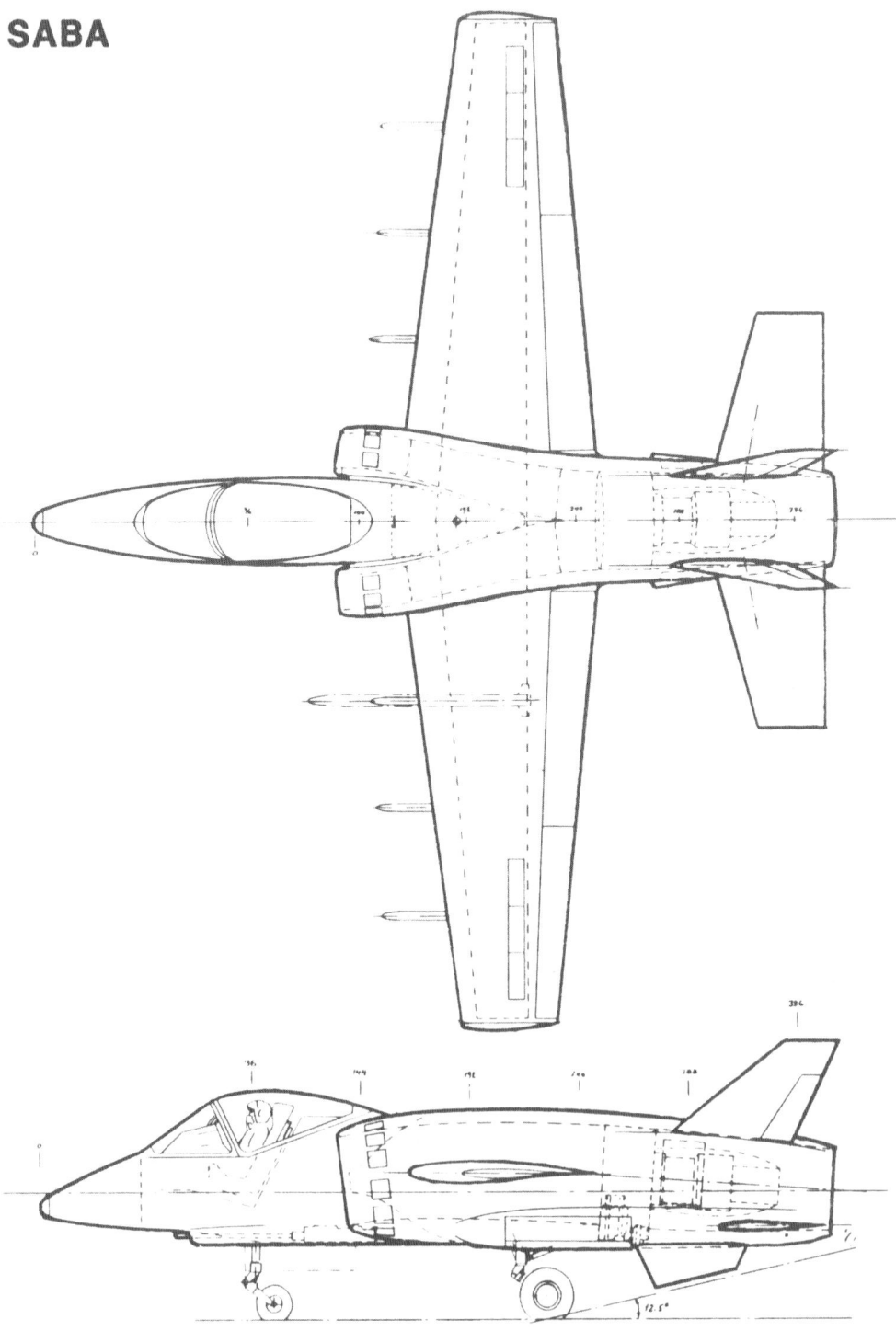

P.1234-2 GA, powered by Avco Lycoming ALF-502 turbofan.

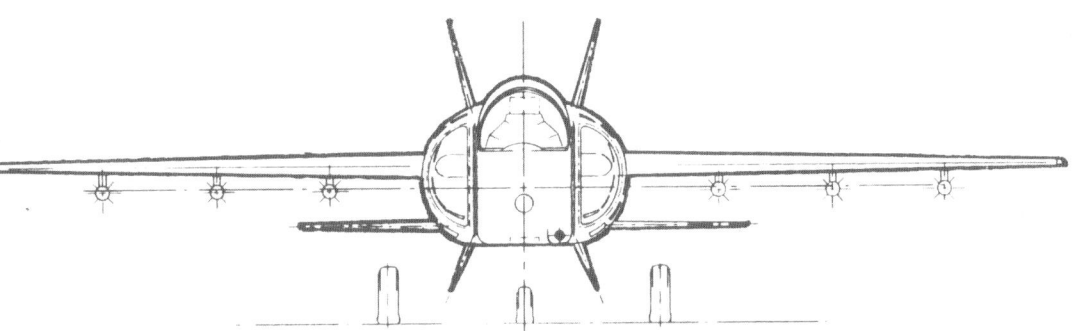

P1234 - 2

Length	9.05m
Span	10.97m
Wing Area	18.58m²
Aspect Ratio	6.5
MTOW	5986kg

Avco Lycoming ALF 502
33.4 kN static thrust

at Weybridge at the end of 1987. The plan for 1988 was to select the aircraft concept by mid-year and achieve an aerodynamic design freeze by December to allow work to proceed towards a demonstrator.

In January 1988 a study was carried out at Warton to assess the SABA concept against alternatives in the anti-helicopter role such as Hawk, PC9D, Harrier, AH-64, XV-15, EFA and MSOW.

As drawn by Miles Huckle on 5 February 1988 P.1233 had been considerably refined but still exhibited the /1 configuration. Wingspan was 10.97 metres with a 13.5-degree leading-edge sweep, giving a wing area of 20.44 square metres and canard span of just over 6 metres. The canopy was so arranged that the front portion was hinged for sideways opening while the rear canopy was hinged to the rear in order to clear the canard in its unpowered position where the trailing edge would project uppermost. However, this was to be superseded by the P.1233/5 of June 1988 which would become the new baseline configuration. The layout replaced the aft fuselage arrangement with twin wing-mounted booms positioned mid semi-span, carrying inward-canted twin fins offering greater protection to the engine and prop. The main undercarriage was now faired into the wing/boom junction. The new configuration sought to answer concerns regarding the stowage of the robust main undercarriage in the wing

SABA

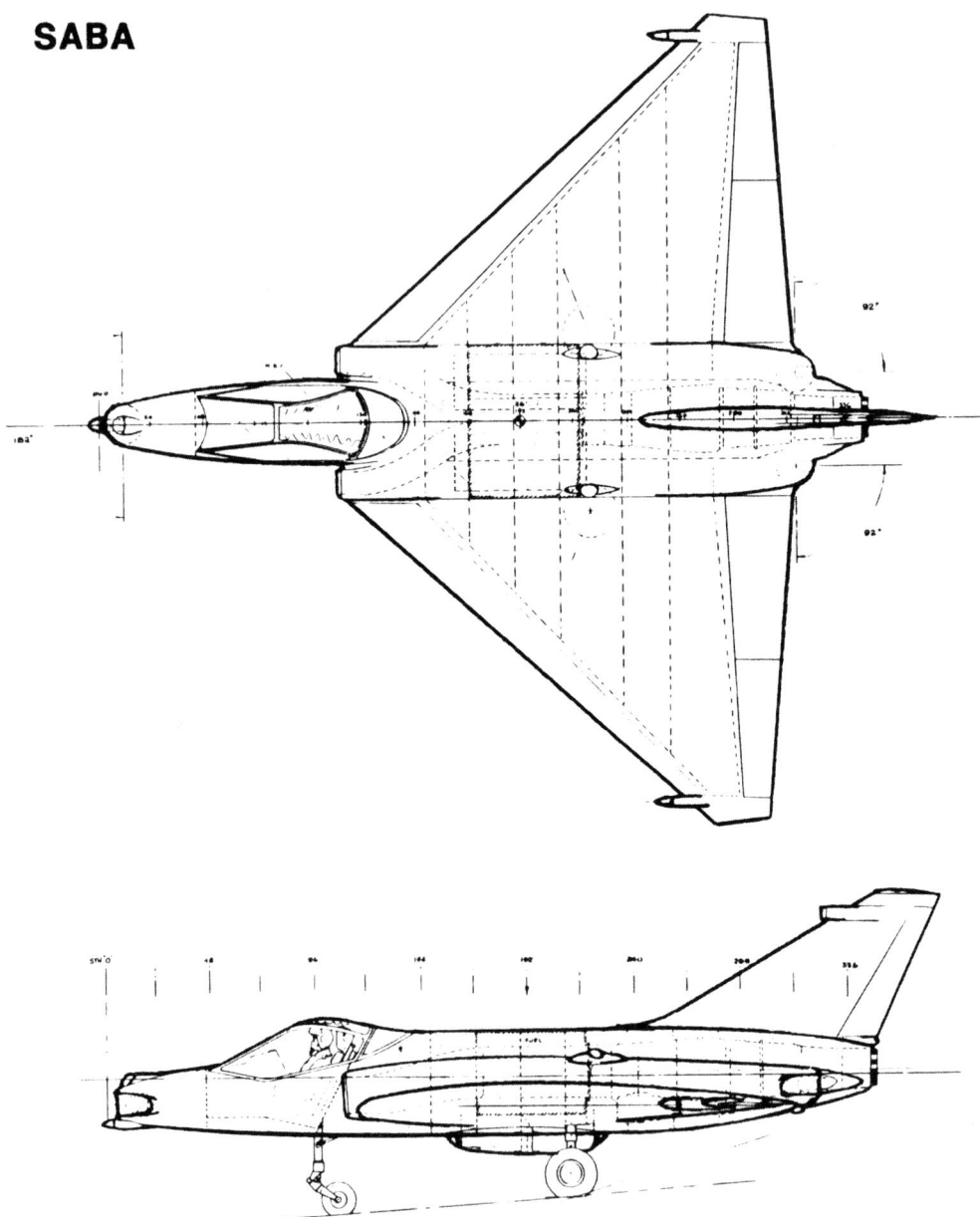

P.1234-3 GA, powered by RR Adour RT-172 turbofan.

and the compromises required in the wing structural box to achieve this. The new layout also eased engine removal by removing the fin structure from the area above the engine/propeller installation.

As the SABA proposal continued to mature, in April 1988 thoughts turned increasingly to the likely market for such an aircraft, especially since this was

SABA: Small Agile Battlefield Aircraft

P1234 - 3

Length	9.53m
Span	9.14m
Wing Area	30.75m²
Aspect Ratio	2.72
MTOW	5782kg

RR Adour RT-172-871
25.4 kN static thrust

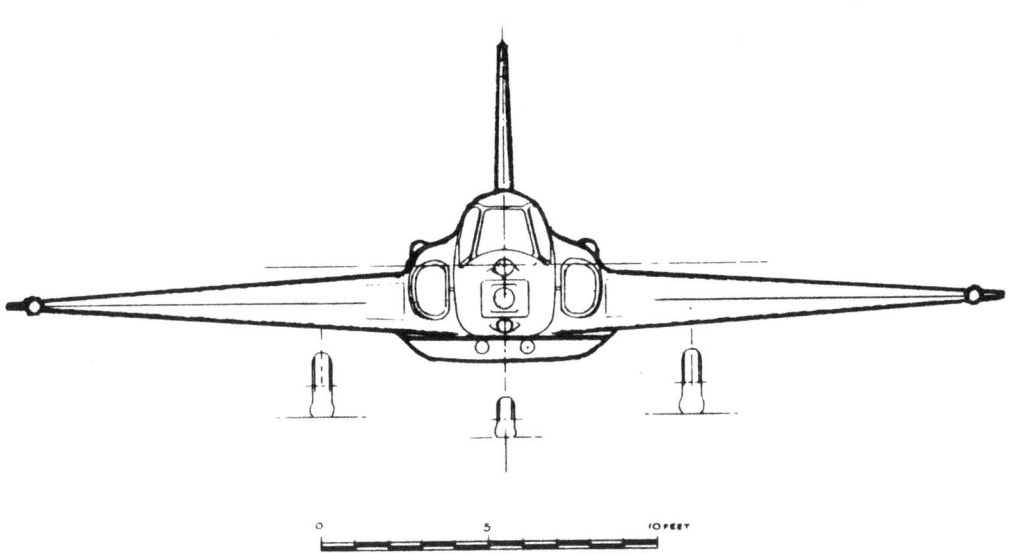

a private venture which had no backing from the UK MoD. The project was aimed principally at the USAF as a close air support aircraft: success there would provide a sizeable launch market, but other markets were also examined, the two most favoured being West Germany where the potential was seen as 100-200 aircraft and South Korea with a possible market of 50-100 aircraft. Other possibilities were China, India or Yugoslavia. Publicity continued to emanate from BAe regarding the project, April 1988 seeing a report published in *Air Force Magazine*, still showing the P.1233/1 layout, shortly to change. The report noted that the company was embarking on an internally funded R&D programme 'in order to validate further the promising results of the work carried out so far ... The funding required has been assessed up to the stage of a demonstrator aircraft, which would make its first flight in 1992 or 1993.' *Jane's* was also keen to place details of the project in their *All the World's Aircraft* annual compendium though the normal requirement to allow inclusion was that metal cutting should have started.[10]

Artist's impression of mixed force of P.1233 SABA and V-22 Osprey tilt-rotor aircraft.

Around June 1988 concerns began to make themselves felt regarding the likely market for SABA as European countries moved towards other, principally helicopter and 'manpads' (man-portable air defence surface-to-air missile system) options for the anti-helicopter role. While the Middle East was seen as a potential area for marketing, concern was expressed that SABA would cut across the market for the Hawk. Keith Towell, now SABA Project Engineer, admitted that the anti-helicopter role was not now seen to be as great as originally thought, the CAS and COIN roles becoming more important for the aircraft. Also, the carriage of ASRAAM missiles, an important part of the role fit, was found to be incompatible with the podded undercarriage and the need to retain a low-drag profile.

Also in June, Andy Jones, now acting in Defence Marketing North America, assessed the project and progress to date, noting 'that the way ahead is far from clear'. He added that 'The concept itself is attempting to break new ground both in technical terms and as a wholly PV military project. The Future Project people are faced with increasing uncertainty (the Kingston office would cease to exist by the end of the year), marketing has been through considerable turbulence and there is a strong likelihood of a major restructuring of the company. There was a row when the project was publicised and now the budget has run out.' Concurrent with this downbeat appraisal were initial costings for the aircraft (i.e., without profit) which foresaw a minimum cost for the AH/CAS aircraft as £6.65 million and a basic 'third world' version as £6 million. This assessment clearly shocked the recipient who noted that Hawk export marks could do the job for less than the no-profit cost of SABA.[11]

In August 1988 a draft five-volume 'Concept Status Report' was produced by Kingston which sought to describe the state of play of the project at that point. Volume 1, the Executive Summary, noted that the primary initiatives were carried out by the Military Division Marketing and Kingston Future Projects teams but, following an internal organisational restructuring, 'a more structured evaluation of the technical feasibility, the operational effectiveness, the market possibilities and business analysis has taken place, including all of the relevant Military Division Functions'. Here then, the project would be exposed to the full weight of company bureaucracy.

Performance was optimised and quantified for the CAS role since this appeared to offer the best chance of a sale to the US Air Force, which had a requirement to replace the Fairchild A-10 Thunderbolt II in the CAS role. Performance goals were seen as high agility – a 180-degree turn in 5 seconds with a 500 feet minimum turn radius; forward basing using rough/soft field performance allowing operation from a 1,000-feet field length; a 400-knot dash speed and 4 hours of loiter time supplemented with 2,500 mile ferry range; all at a target cost of $10 million per unit. The study was based on the datum aircraft P.1233/5 configuration with twin booms offering a wingspan of 10.97 metres and area of 17.7 square metres, a fuselage length of 9.98 metres, a design take-off weight in the order of 6.8 tonnes, powered by typically, a 5,000shp prop-fan.

Notwithstanding the August report, by the summer of 1988 the status of the project seemed less than clear, even to its adherents. On 31 July 1988 P.W. Liddell,

Head of Advanced Studies Department at Warton, in a letter to Mick Mansell, Director of Technology, noted that he had been approached by the editor of *Defence Helicopter World* 'trying to assess the current state of the programme. Thus far I have merely pointed out that, whilst not dead, the programme is currently the subject of review to determine whether or not a sensible route forward is apparent. Given differing signals from various of our directors, covering the spectrum between following the business case recommendations, i.e., to review the anti-helicopter concept, and "killing" the programme, some clarification would appear to be necessary ASAP.'[12]

On 17 August Andy Jones responded to Mike Turner, 'You asked if I still believe in this project? As far as the present design is concerned, I cannot answer yet because very little assessment has been done. We don't yet know whether the performance goals can be achieved, nor is there any real indication of cost.'

At the SABA progress meeting in October 1988, the Concept Status Report noted above was discussed; the Executive Summary (Volume 1) and Business Case (Volume 5) being to hand, these volumes provided the main substance for deciding on the way ahead for SABA.

1) The main deductions from the report were that the concept was technically feasible and that about 1,000 sales would be required to achieve an IRR of 20%.
2) SABA was conceived as a high agility anti-helicopter aircraft, but has subsequently become a dual role aircraft because of the better market potential in the CAS field. Consequently, the project is not optimised for the latter role, although the extra agility may be a desirable feature.
3) The market forecast is critical to the SABA case … the North American market in particular appears to be vital.
4) Operational analysis studies suggest that there is some conflict between SABA and Hawk in the major roles.[13]

In plain language, SABA was facing an uphill struggle: the project would only enter profit after 1,000 sales, a steep requirement; the aircraft was to be offered in a role for which it was not optimised; if the US market could not be penetrated, volume sales were unlikely; and, lastly, conflict was apparent between SABA and the company's profitable Hawk business.

By 11 October Andy Jones was feeling even less positive than in August. He reported to Len Milsom, 'I feel that the project has gone off the rails. The crucial issue is the size of the anti-helicopter market. If that market does not warrant the development with unique capabilities, then so be it … The last thing we should do is pour development money into yet another CAS aircraft – we have enough already and the market is overcrowded – and the likelihood of the USA being a launch market on the basis of the CAS requirement is very remote.'[14]

On the 17 October Milsom responded, with regard to the review, now complete, stating 'The original intention was for the aircraft to be small with a combat weight of about 10,000lbs and address the anti-helicopter role with an emphasis on agility. The market would not interfere with that for Hawk or other BAe products. During the review, the size of the anti-helicopter market was seen to be fairly small while the potential for the CAS role appeared and work was done to try and meet this requirement. This resulted in a heavier aircraft which then began to seriously overlap the potential of the Hawk. Furthermore, to launch brand new aircraft of this type would need a market of 1,000 aircraft in order to make a reasonable return and the US CAS market was seen as the only real possibility.'

'But what of the future?' Milsom pondered. In essence, he was admitting that the project was at an end. Modest funding would be authorised till the end of the year with an intention to 'take stock in the New Year to determine whether any further work beyond what I have outlined is then justified'.[15]

However, Jones, now Vice President, Defence Marketing (North America), was not yet ready to capitulate. In seeking a turn away from the business case model and a return to his original concept he suggested,

1) SABA should be reassessed in terms of the market for an anti-helicopter variant.
2) It should be assumed that the original targets for weight, speed, performance, agility and rough field capability can be met.
3) The cost of this variant should be assumed to be two-thirds of the variant in the business case (being two-thirds the weight).

   If as suggested we need 1,000 sales of a variant largely designed for the CAS role, it would be foolhardy to continue. If, however, we could anticipate a profit from 300 or so sales of a version which, in its primary role, does not conflict with any other of our products there may yet be a case worth pursuing.[16]

Following his retirement, Ralph Hooper remained lukewarm about what he called Jones' 'chopper-chopper'. In a letter to Mike Turner, Hooper, while appreciating how difficult a target a low-flying helicopter could be, felt that the company needed to garner full-scale experience of the anti-helicopter role before formulating too detailed a response. 'Surely there is some existing aircraft that can be modified (more thrust? Bigger wing?) to yield turn rates approaching those proposed and could be used as a one-off trials aircraft? A higher fixed wing turn rate obviously would help but can it be conclusive? If a hunter-killer submarine is the best way to deal with subs, perhaps we need a hunter killer helicopter.'[17]

By November 1988 the work of the Future Project Office at Kingston was winding down, its role as an independent project office at an end after nearly fifty years. SABA activity was likewise winding down, and also transferring to Warton, a black hole drawing all activity within BAe into its maw.

Why did this promising project fail to achieve flight status? One might point to the lack of a suitably sized market for what could be considered a niche product. Certainly, without a US order, there was never any realistic likelihood of the aircraft going into production. Also weighing on the project was the inevitable growth in size and weight, together with concerns regarding survivability in the hostile environment that FEBA (Forward Edge of Battle Area) presented. Although aircraft such as the Sukhoi S-25 Frogfoot and the Fairchild A-10 Thunderbolt II, for which SABA was suggested as a replacement, had shown that hardened close air support aircraft had their place on the battlefield, subsequent development of portable anti-air and anti-tank weapons deployed by infantry have reduced the need for such a platform. It is perhaps pertinent here to note that the A-10 was designed as an anti-tank platform rather than anti-helicopter, which was the original role envisaged for SABA.

Internally, problems lay with the legacy of the original BAe amalgamation which had left two competing design houses, Kingston and Warton, vying for projects and finance to develop them. With the creeping takeover of senior management by those whose loyalties lay in the north, Kingston was finding it increasingly difficult to obtain funding agreement for SABA and indeed for their other big project, the ASTOVL P.1216 of which, more in the next chapter. Ultimately, the drift of influence to the north saw the determination to achieve the closure of the southern sites, Kingston and Dunsfold, only increase.

Chapter 7

ASTOVL: The Supersonic Ambition

Although Hawker Siddeley's P.1154 supersonic Harrier proposal had foundered in the financial quagmire that the UK found itself in, in the 1960s, the company remained convinced, as did Rolls-Royce Bristol, that a supersonic V/STOL aircraft was not only possible but indeed distinctly probable within the next ten years. By 1969 the UK was operating the world's first V/STOL military aircraft in its frontline inventory; HSA was aware that it led the world in V/STOL technology and was cognisant that, if it could only develop a better mousetrap in the form of a supersonic Harrier, the world would beat a path to its door.

Over the next twenty years, HSA Kingston Future Projects would design over sixty-two V/STOL proposals (63 per cent of total HSA/BAe Kingston proposals between 1960 and 1987) in an effort to profit from their lead in V/STOL technology; ultimately it would fail to do so, while the US would eventually steal its thunder with the F-35B in the twenty-first century. That it would take a full fifty years before a supersonic ASTOVL combat aircraft entered the inventory of the US and UK perhaps reflects the difficulties that HSA, and later BAe and BAE Systems, would have faced if a UK ASTOVL project had been fruitful.

Following the cancellation of P.1154 in 1965 much of Kingston Future Projects work revolved around producing an acceptable V/STOL aircraft with minimal changes from the P.1127 format for service entry in 1969 as the Harrier. In the event, significant changes were made, and Harrier was able to enter service on time and become an effective ground-attack aircraft capable of basing close to the front line in Europe. Thus, it was not until 1967 that thoughts again turned to supersonic V/STOL or, as it would come to be known, ASTOVL, advanced short take-off and vertical landing. The P.1175 was a step in this direction, though still subsonic, featuring a Pegasus development engine and an Allison lift jet, this project growing into the P.1177, a similar layout but with supersonic capability powered by a Pegasus BS.53/103 vectored-thrust engine with reheat and an additional Rolls-Royce/Allison XJ-99-RA-1 lift engine. A

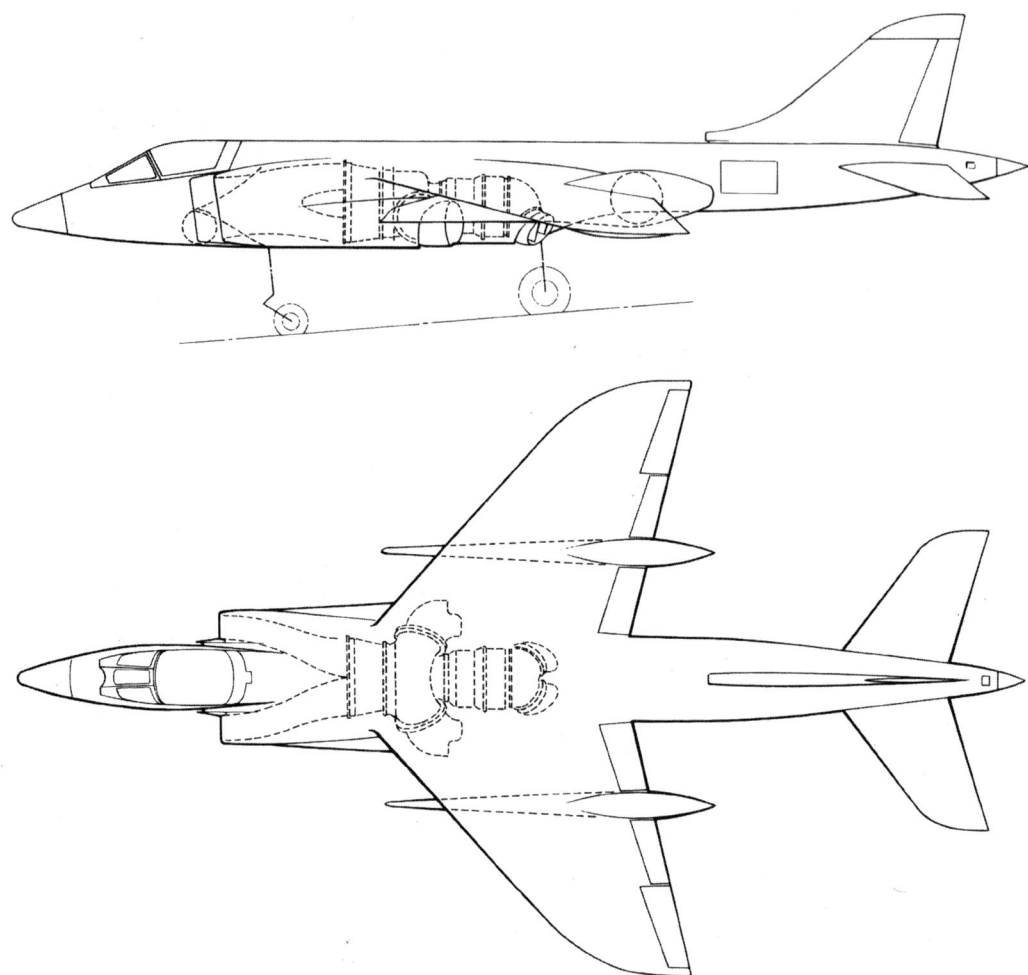

P.1179L GA, renewed thoughts on supersonic V/STOL.

variant powered by twin Rolls-Royce Spey 5R engines plus lift engine or Rolls-Royce RB.168-31D (but without the lift engine) in place of the Pegasus was also drawn. These designs featured nozzles drooped and trailed to reduce the impact of the jet thrust on the rear fuselage when in cruise mode.

These format studies were continued in the later 1960s and early 1970s with P.1179, designs being produced in response to calls for a Multi-Role Combat Aircraft for NATO. The designs would follow the general Harrier format but with a single or twin Rolls-Royce RB-199 engine/s featuring drooped and trailed nozzles with PCB and reheat or developed RR Pegasus 15 or 9 engine, again with PCB. At this time, variable geometry (VG) wings were being discussed and drawn in many of the UK design offices, largely driven by work carried out

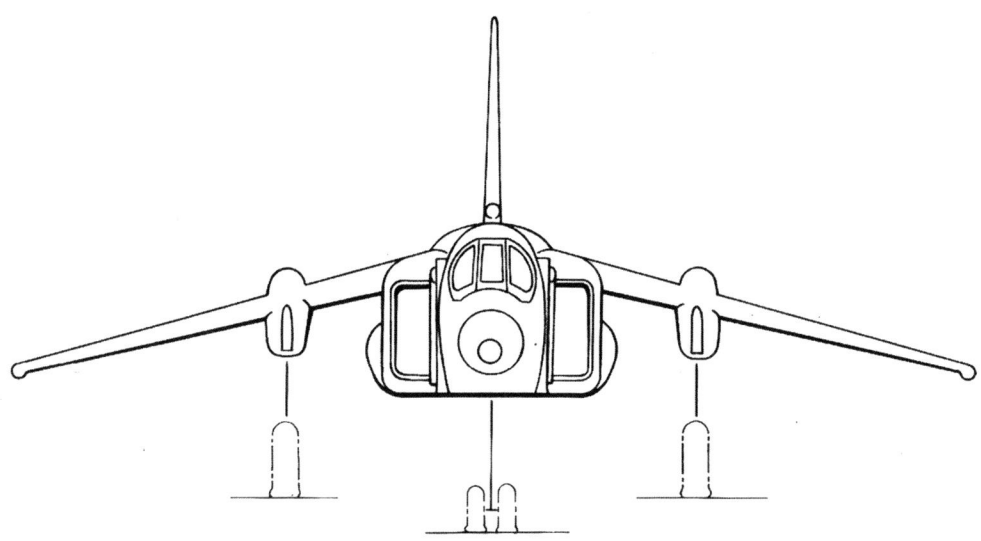

R.R. Pegasus 15 Engine with P.C.B.

by Sir Barnes Wallis at Weybridge, HSA being no exception and two drawings, P.1179P and P.1179U, show work by Brough to produce a VG variant under this project code powered by conventionally arranged (i.e., not V/STOL) twin RB.199 engines. Ultimately, the successful tender for this requirement was awarded to Panavia Aircraft GmbH, a European partnership between BAC in the UK, West Germany and Italy, leading to the Tornado strike aircraft with twin RB.199 engines and variable geometry wings.

In the early 1970s a requirement, AST.396, for a replacement aircraft for Harrier and Jaguar was raised by UK MoD. For HSA Kingston, what could be a better Harrier replacement than a better Harrier. Design work was now aligned with the AST requirements and coded HS.1184, initial designs in 1971 being drawn around the Pegasus 15 engine with and without a separate lift fan behind the cockpit. One frankly bizarre design, HS.1184-16, featured a second cockpit at the rear of the aircraft and rear-firing missiles together with twin canted fins. By the mid-1970s this work had begun to align with thinking at MDC (McDonnell Douglas) in the US in their own search for an advanced Harrier for the USMC. This, the AV-16S, would feature a developed VT (vectored-thrust) engine with PCB, and much greater range and load carrying, the hope being

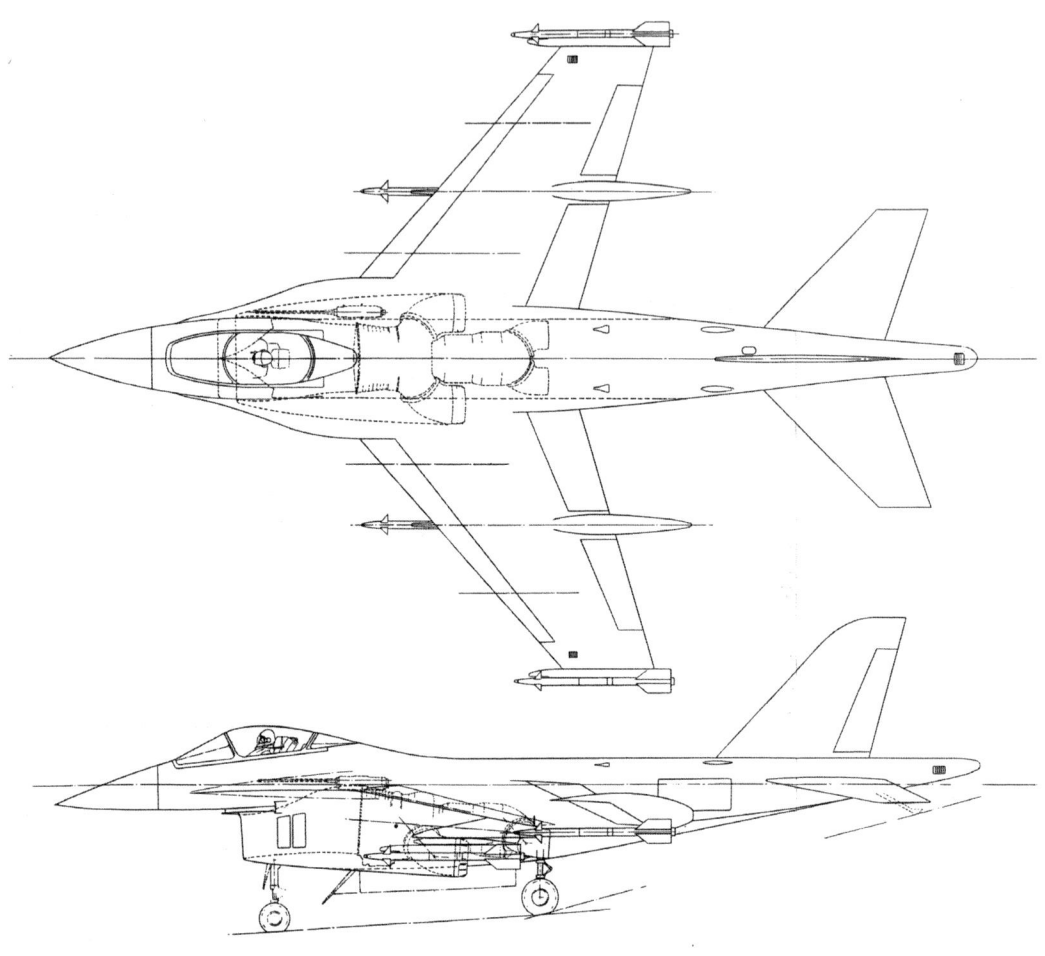

P.1205-9 GA, supersonic STOVL Harrier development.

that such a design might be attractive to the UK as well as the US, especially since development costs could be shared. Now coded HS.1185, design work continued with MDC in an effort to reach a design that would be acceptable to the US Navy, under which USMC operated financially, as well as the Marine Corps. Further work proceeded via HS.1186 and HS.1187, some designs featuring innovative elements but mostly still based on the proven VT four-poster technology. Ultimately, a developed Harrier under AV-16 and HS.1184 depended on the development of the Pegasus 15 engine jointly between Rolls-Royce and Pratt & Whitney, for which substantial funds would be required. When, in the early 1970s, it became clear that funding would not be available,

ASTOVL: The Supersonic Ambition

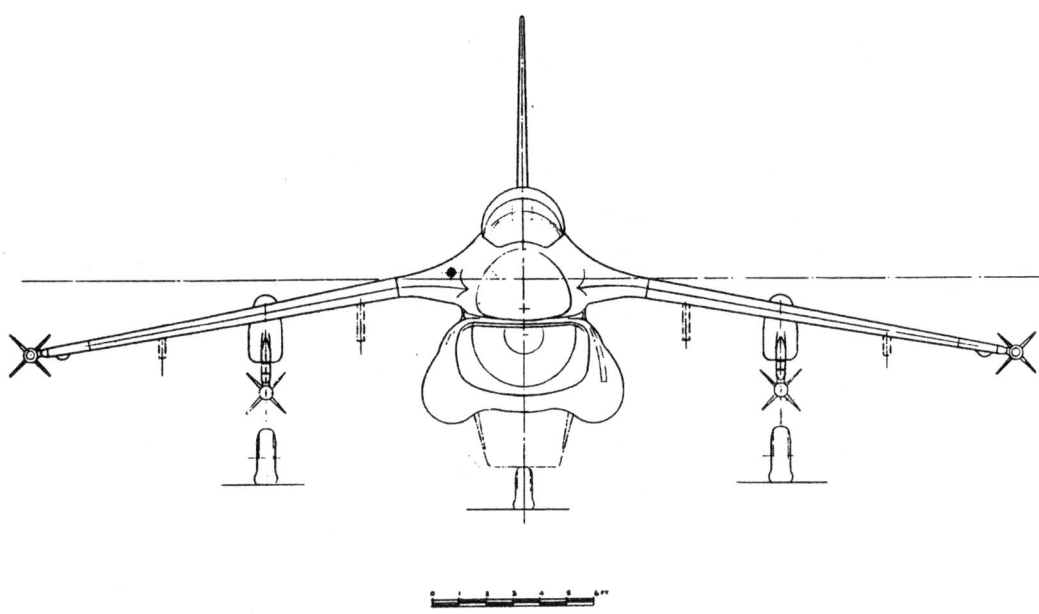

McDonnell Douglas, Hawker Siddeley's partner in the US, was asked what could be done to enhance the aircraft using the existing engine, these thoughts eventually coalescing around what became the AV-8B.[1]

HS.1205

By 1976, having in the meantime covered other design formats covering standard engine layout strike fighters and launched the remarkably successful Hawk series of aircraft, thoughts at Kingston Future Projects had turned once again to a supersonic V/STOL aircraft. HS.1205 once again started as a basic Harrier four-poster Pegasus layout with drooped and trailed nozzles, PCB and podded undercarriage, Vic Hancock creating the first drawing. This was then modified by dropping the engine thrust line down below the aircraft CG, partly to give greater fuselage space for fuel or equipment, but also in an attempt to alleviate the impact of the jet exhaust on the fuselage structure. This last point was an important one; one of the Harrier's shortcomings was the fatigue induced in the rear fuselage by the jet plume impacting on the structure in flight, leading to acoustic fatigue and increased tailplane vibration, to the detriment of both tail and avionics. Dropping the thrust centre allowed the jet plume to exhaust below the fuselage with less impact on structure. However, this remedy could only be taken so far. The divergence of thrust centre and CG would lead to pitch control

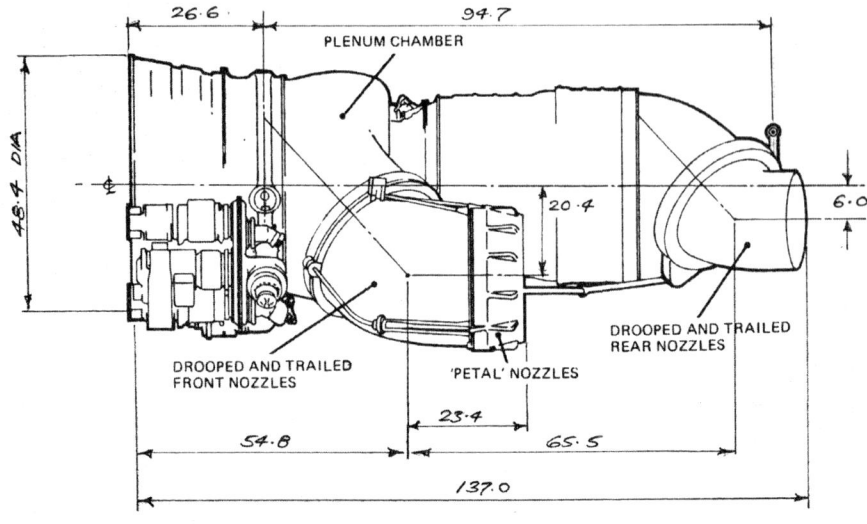

Pegasus 11F-43 engine proposed for P.1205.

problems during the transition from vertical to horizontal flight, part of the flight regime that certainly did not need additional control difficulties.[2]

In the mid-1970s, AST.403, a re-issue of AST.396, was sent out, again calling for a replacement for Harrier and Jaguar. In response, Miles Huckle's HS.1205-5 of December 1976 featured a chin air intake feeding a Pegasus 11 with PCB, wing podded undercarriage and all-round pilot visibility with semi-supine pilot seating and mildly swept wing with small Leading Edge Root Extensions (LERX) but, by October 1977, Vic Hancock's drawing shows separate low-slung intakes at the 4 and 8 o'clock positions, wing fold and twin canted fins. However, the engine installation was still entirely housed under the main fuselage. As the design progressed, the chin intake re-appeared and, at this point, probably around 1979, alternative layouts, HS.1205-25 and HS.1205-26, featured the rear fuselage split into two and a single rear nozzle exhausting in the space created, an innovation first schemed on the P.1187 in 1970 which would feature large in subsequent efforts to achieve a true ASTOVL fighter. However, principally due to insurmountable pitch control difficulties during transition encountered due to the low position of the nozzles, work on this project ceased in 1979, studies moving on to the P.1212.

P.1212

The P.1212 was a significant step on the way to BAe Future Projects Kingston preferred solution to a workable ASTOVL design. Most of the studies under this project code featured the split rear fuselage as witnessed on the first variant. This featured what might be termed a notched delta wing carrying the fuselage aft via two booms mounted at mid semi-span, the wing surface inboard of these being cut back to create a space through which the engine rear nozzle could propel the aircraft. Two further nozzles mounted to port and starboard on the approximate CG completed what was a three-nozzle vectoring powerplant having the minimum of acoustic and vibration impact on the rear fuselage. No tailplane was carried but twin canted fins were drawn at the rear of each fuselage pod. The engine was fed via a chin air intake and the engine location was again below the main fuselage CG datum, promising problems with flight transition from the hover to forward flight and back again. Various wing layouts were considered including moving forward the wing portion outboard of the booms (P.1212-3), sweeping forward this same portion creating a W planform (-5), a 'standard' 40- or 45-degree sweep back (-6 and -7) and reverse wing taper (-10), (with the greatest chord at the tip instead of the root). However, it was the standard swept-back wing that was favoured, this carrying the fuselage booms at mid semi-span which in turn carried the fins. For this layout tailplanes were added on the outboard of each pod, leaving the inner space clear for the jet efflux. This stage was reached in September 1979. However, as with the P.1205, control problems led to the abandonment of the design.

For a time, P.1212 was seen as a most promising layout, to the extent that a flying demonstrator was proposed to gather further information on the concept. Two MoD PE study contracts were raised against which BAe defined what parameters might be most usefully investigated using a demonstrator. These covered the conventional V/STOL flight characteristics up to Mach 1.5 using a Pegasus 11 engine modified with PCB in a three-nozzle layout. A decision to go ahead would have likely seen a P.1212 first flight in 1985, with development costs for two aircraft being in the range of £50 million plus the associated engine costs. Unfortunately, the proposal went no further.

Before moving on to the definitive layout for Kingston's ASTOVL aircraft, two further projects looked at other novel wing layouts. The P.1213 returned to a single fuselage and swept-wing planform, but with a large canard foreplane with dihedral while initial P.1214 studies used a forward-swept wing as well as the canard layout. The P.1214-3 dispensed with the canard and instead adopted

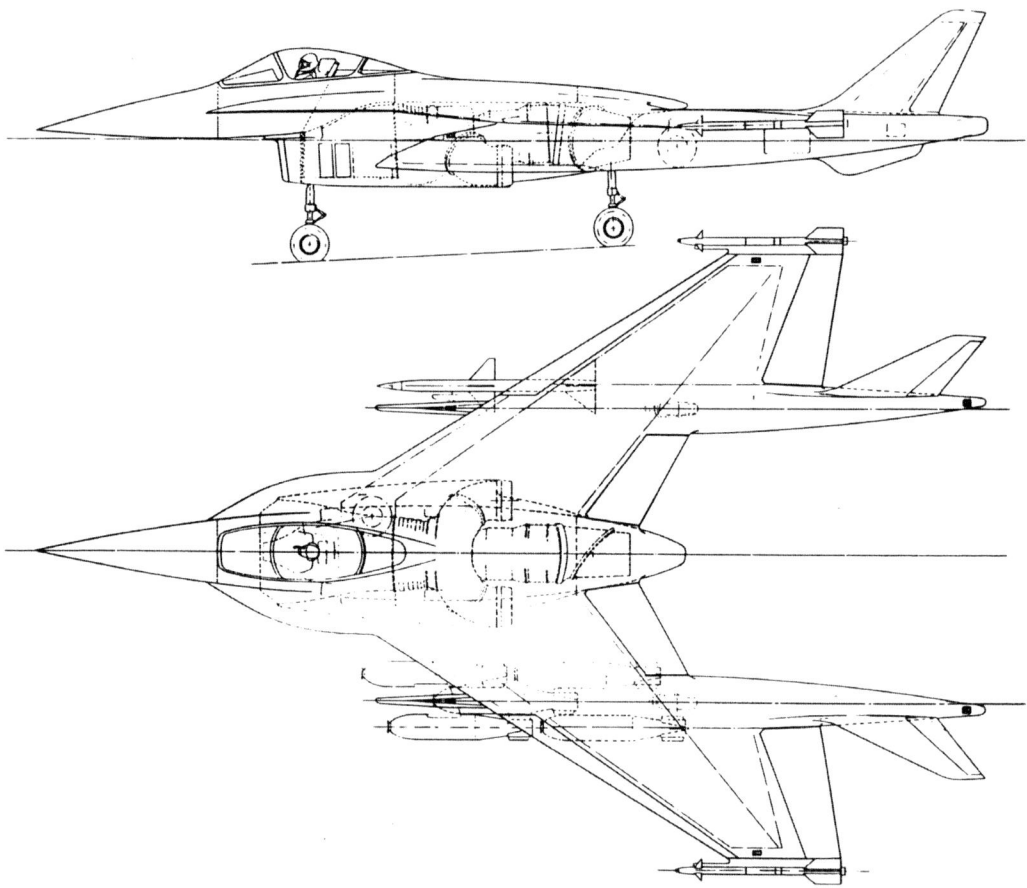

P.1212-2C GA ASTOVL aircraft featuring twin booms.

an X-configuration with forward-swept wing and conventionally swept tail surfaces, but again freeing the rear space for the single-nozzle jet efflux.

P.1216

Work on P.1216-1 began in the late 1970s. As originally conceived, it owed much to the P.1212 studies, being similar in layout. That layout comprised a wing swept at approximately 50 degrees supporting at mid semi-span deep twin booms carrying a fin and tailplane together with a vestigial ventral fin. The shoulder-mounted wing exhibited moderate anhedral and carried two thirds of the fuel. The short fuselage was blended into the wing via LERX below the cockpit. Initial thoughts on a suitable engine were based around the Rolls-Royce RB.422-48, fed via a substantial chin intake, supplemented by blow-in

ASTOVL: The Supersonic Ambition

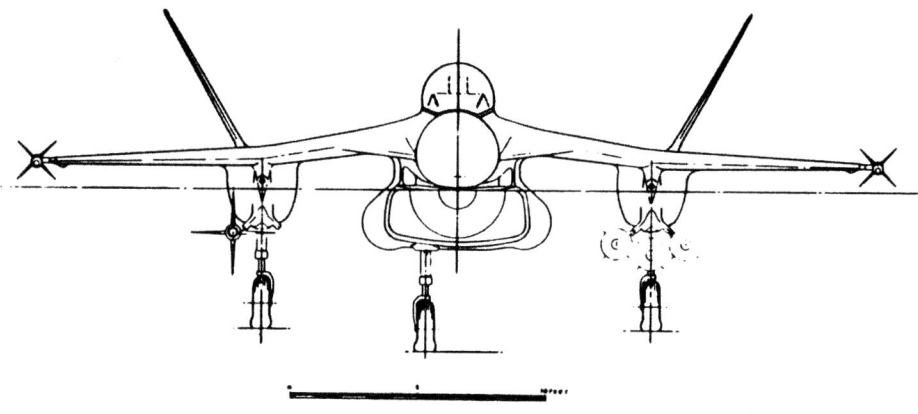

P.1212-2C WITH PEGASUS 11F-38B ENGINE

doors around its periphery, and exhausted via a main central nozzle directly behind the engine on the centre-line and twin side-mounted drooped nozzles a short distance forward, all fully vectoring. Predicted power was not stated but presumably reheat or PCB was envisaged in the package. Armament consisted of missiles mounted at the wing-tips and either missiles or bombs carried under the booms, depending on role. The principal difference to the P.1212 involved a lengthening of the booms and the addition of outboard tailplanes to improve the control arm.

The refined P.1216-2D as drawn by Vic Hancock featured a Rolls-Royce Pegasus 11-03B engine and modified auxiliary inlet doors to the intake but otherwise was externally little changed, while the P.1216-3 saw an increase in fuselage length and the guns moved from the front of the booms to locations in the intake wall under the LERX/chine. At this stage the dimensions of the proposal were somewhat larger than the Harrier with fuselage length, i.e., including booms, of 57 feet, wingspan of 32 feet and area 427 square feet. The design would continue to be refined through many iterations from 1980 through to 1988, some fifty different variants being produced, but these would involve fairly minor adjustments to the basic P.1216 layout, the twin-boom design being almost constant throughout.

Minor digressions from the basic layout included the 'cranking' of the wing leading edge on the P.1216-8 version whereby the sweep was reduced outboard of the boom locations and the P.1216-9 which featured variable-geometry wings which would sweep outboard of the booms (one wonders how the reaction

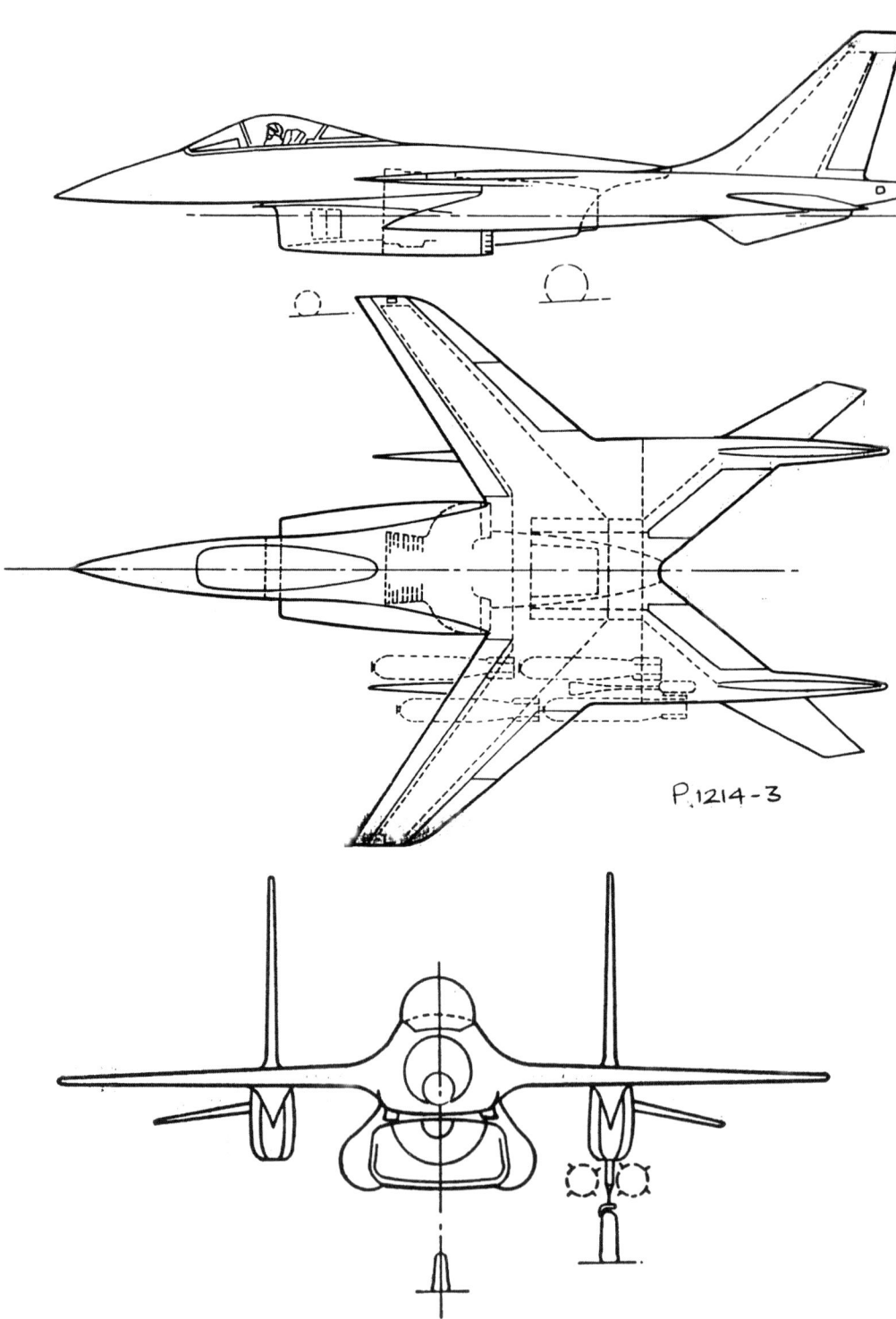

P.1214-3 GA forward-swept ASTOVL layout.

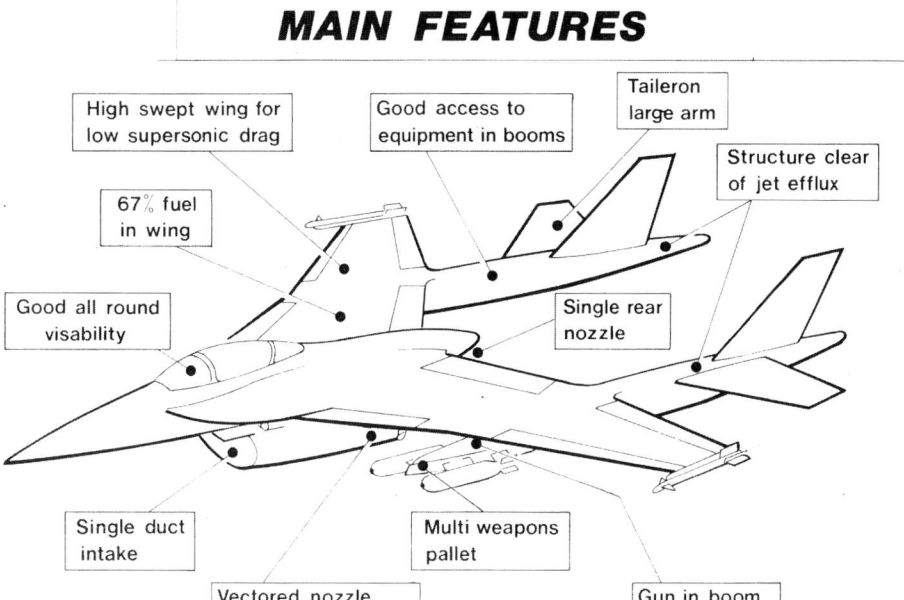

P.1216 ASTOVL aircraft main features.

control 'piping' would have been led through the swivel joint); these were, however, not proceeded with, the wing retaining a conventional moderate sweep in later proposals. The Kingston staff felt that one of the primary advantages of its somewhat unconventional rear fuselage arrangement was the easy access to avionic equipment which would be located in the tail booms well away from the exhaust plume. Whether this would have been the case is moot; certainly, Harrier's easy avionic accessibility would take some beating.

Since Kingston was seeking to offer the design as a combined air combat and ground-attack aircraft, effort was expended in ensuring that the design was optimised for both roles. In the P.1216-13, armament for the air combat role was seen as comprising a suitable search radar coupled to four AMRAAMs for long-range interdiction and two ASRAAMs for close-in engagements. For the ground attack role, a mix of BL-755 cluster bombs and two ASRAAMS together with two 270-gallon fuel tanks was drawn, the tanks being carried outboard of the booms. A naval variant, drawn as either single- or twin-seat variants with wing folding, saw the surface attack role replacing the BL-755 with four Sea Eagle anti-ship missiles while for the CAP role a combination of four AMRAAMs, two ASRAAMs and two 370-gallon fuel tanks would be

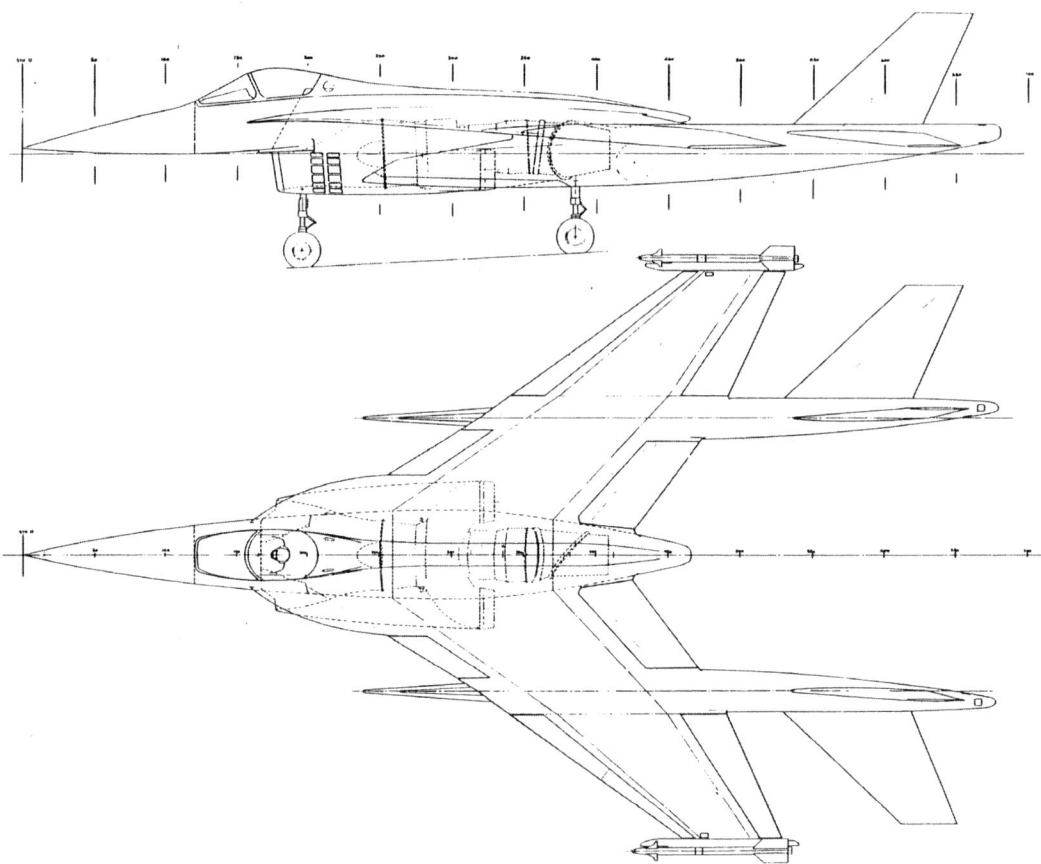

P.1216-2D GA with Pegasus 11-03B engine.

carried. These studies would become important later when a replacement for the Sea Harrier was being discussed as this aircraft was entering service.

By the time that the P.1216 had reached the fortieth variant, detailed work on the flying controls had resulted from exacting manoeuvrability requirements, in particular turn radius. The P.1216-43 for example, featured leading-edge flaps for pitch trim, all-moving fins for yaw control, tailplane operation coupled to inboard and outboard flaps (symmetric operation for pitch control or asymmetric for roll control). The final drawing variant appears to have been the P.1216-50. Thereafter work ceased with the removal of Future Projects from the Kingston/Weybridge enclave, all future crystal-ball gazing moving to Warton.

Most of the later P.1216 iterations were projected to be powered by the Rolls-Royce RB.422-60 which was expected to offer up to 40,000lb thrust, promising speeds of up to Mach 2. This would be coupled with a three-nozzle exhaust, the front two having variable nozzles and fitted with PCB. Detail work to make

ASTOVL: The Supersonic Ambition 163

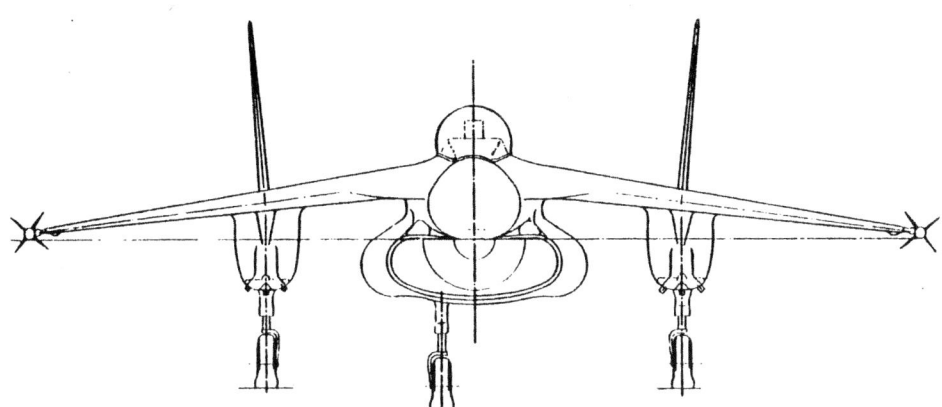

engine change as simple as possible (the Harrier engine change required the wing to be removed) resulted in projected engine change time being as little as fifteen minutes by two engineers, the ECU being dropped out of the bottom of the fuselage directly onto its handling trolley. It would have been most

P.1216 powered model for investigating hot gas re-ingestion effects.

interesting to see this put to the test. The RB.422-60 was conceived at Rolls-Royce Bristol in 1980 as an all-new replacement for the Pegasus and specifically for the P.1216, though with applications to future re-engining of the AV-8B but did not proceed further than parametric studies. A much higher fan pressure ratio, up to 1:4, was planned for the P.1216-60 but an increase in front-nozzle thrust was offset by lower bypass ratio allowing more air into the main core for combustion, thus increasing the thrust available to the rear nozzle. In the event, as with P.1216, the engine did not reach manufacturing stage. However, trials of the PCB concept were re-started at Shoeburyness in 1983. These had first been undertaken during the days of the P.1154 and the BS.100 engine at Anstey in 1964. For the later series of tests, model trials were superseded by full-scale trials using a modified Pegasus 2 fitted into a composite Harrier airframe which was suspended on a large gantry that in effect allowed the aircraft to 'fly' in order to assess the effects of hot gas re-ingestion.[3]

Attempts to control the weight of the aircraft led to extensive studies in the use of carbon-fibre composites, mainly in the wing. At the time this was something of a new technology, although it featured large in the AV-8B/GR.5 Harrier, and there were concerns regarding the stiffness of the resulting structure. In particular, there was widespread concern that the long booms mounted on the wings would result in unacceptable aeroelasticity – bending of the wing under load with consequent loss of control. Aware of the risks that this configuration posed, Kingston staff were able to tailor the wing and boom

P.1216 low-speed wind-tunnel model.

assemblies such that the tail provided 'an amplifying effect to the action of the outer wing ailerons, in which the tail reacted against the wake of the aileron to change its own lift, twisting the boom and inner wing. Thus ... the elasticity of the wing was turned from a weakness into an advantage'. Any induced bending would in fact now work with the flying controls rather than against them, giving greater control.[4]

However, the weight issue would not go away and became more pressing when the Kingston and Warton Project Offices were both working to produce their respective designs for submission as a package to the MoD. Despite Kingston's best efforts under Head of Future Projects Chris Hansford, Ralph Hooper, by then Divisional Technical Director, was not at all happy with the resulting weight penalty that the aircraft was carrying. In one of his many handwritten memos, Hooper did not 'do' computers, he was adamant that the Project Office must do better. In comparing Kingston's offering against the Warton studies Hooper in May 1984, blasted 'The question remains the same – you have wiped out a 10% engine weight advantage and a 100% engine volume advantage and come up with inferior performance (even with the bigger wing) and without claiming any countervailing performance advantages of your choosing ... You have suggested that the P.112 (one of the Warton offerings) is mildly volume deficient; I offered the excuse that the P.1216 is always carrying its "pylons" ... and we would accept some wetted area penalty from the twin rear fuselages of P.1216 but these scarcely seem to offer a satisfying explanation for the MD – who is about to visit south and will doubtless be fed the merits of 279 (the MDC VT proposal). WHAT DO YOU HAVE TO SAY?'[5]

Hansford had to admit that Hooper's points were correct so far as they went. However, 'The advantages come from reduced in-service costs (more reliable, easier maintenance etc.), and to my mind are sufficiently large to justify the layout, despite a small performance *disadvantage*. Relative to the P.112, there are also risk comparisons that must be made, and judgements as to the value of STOVL as opposed to CTOVL; also, the integrity hazard posed by the flow switching implications. Most of these operate in our favour, but are not quantifiable in the sort of comparison that you have asked for. I would like to give the MD a rosier picture but only if based on reality.'[6]

Given the hugely expensive costs of investigating the provision of state-of-the-art aircraft in the current market, BAe, like other aerospace companies, sought to minimize their PV (private venture) work by seeking MoD contracts for study of RAF requirements. In 1982 the requirement exercising the RAF

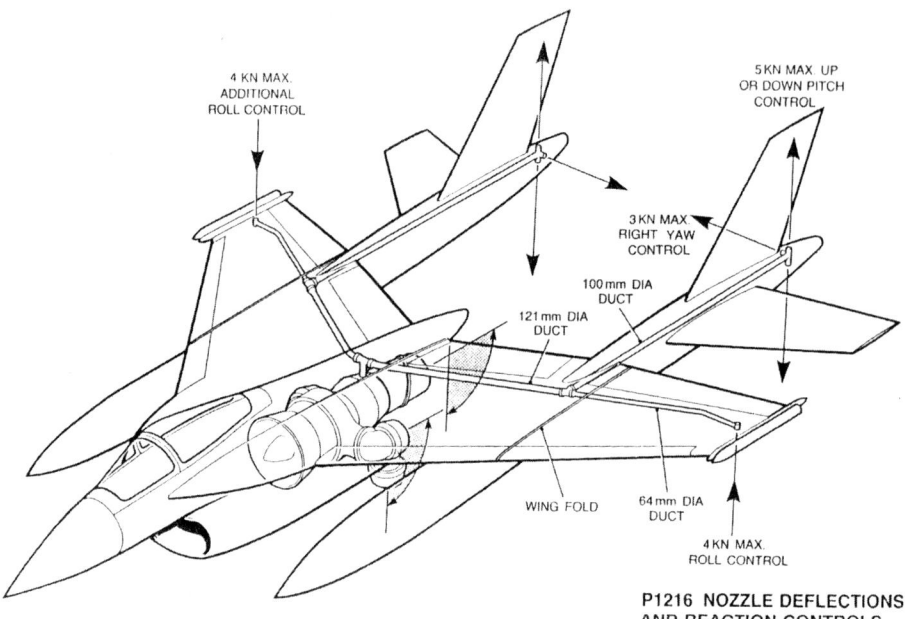

P.1216 nozzle deflection and reaction control schematic.

was AST.410 to replace the Harrier (GR.5) and the Jaguar and, later, in 1985, NST.6464 to replace the Sea Harrier in the Royal Navy.

At the same time, Rolls-Royce and BAe (Kingston and Warton) were examining other jet-lift concepts, these being tandem fan (P.115), RALS (P.112) and ejector lift (P.116). These studies would later read across to the Royal Navy requirement for a Harrier replacement and be the cause of much bad feeling between the Kingston and Warton design offices.

The work within BAe on the Royal Navy's NST.6464 was split into two stages at the request of the MoD which saw the likelihood of obtaining an ASTOVL aircraft anytime soon to be remote. Study A was to submit an interim design that could replace the Sea Harrier in the short term, i.e., by 1997, while Study B would propose designs rather more advanced, investigating other STOVL type powerplant arrangements as noted above. Study A, based only upon VT aircraft, looked at using a P.1216 derivative from Kingston, P.1216-41, and a P.109 from Warton, management authority residing at Kingston. Other possibilities were the MDC 279-3 and developments of the FA.2 Sea Harrier and the GR.5 Harrier. Study B included a developed P.1216 (-43) from Kingston featuring vectored thrust and three designs from Warton, P.112 using RALS, P.115 using tandem fan and P.103 using tilting nacelles, management responsibility

ASTOVL: The Supersonic Ambition   167

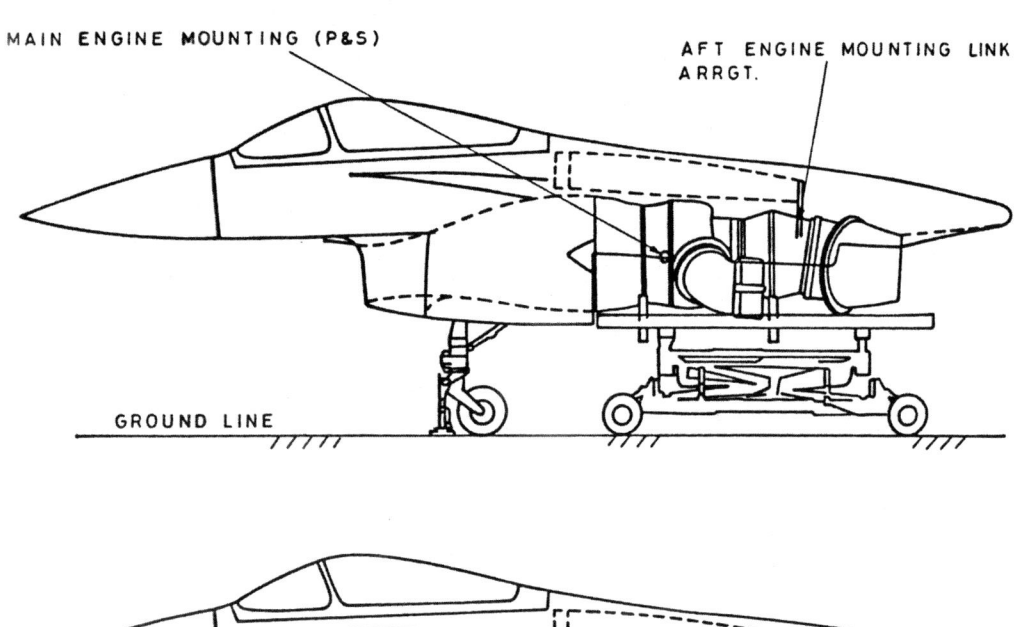

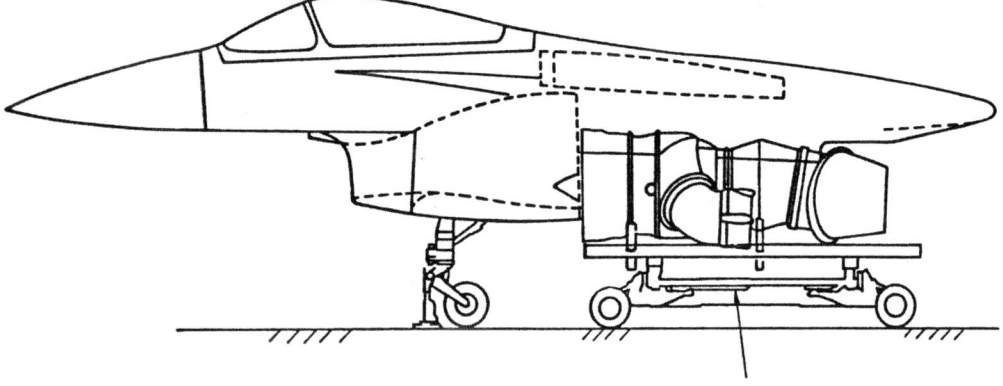

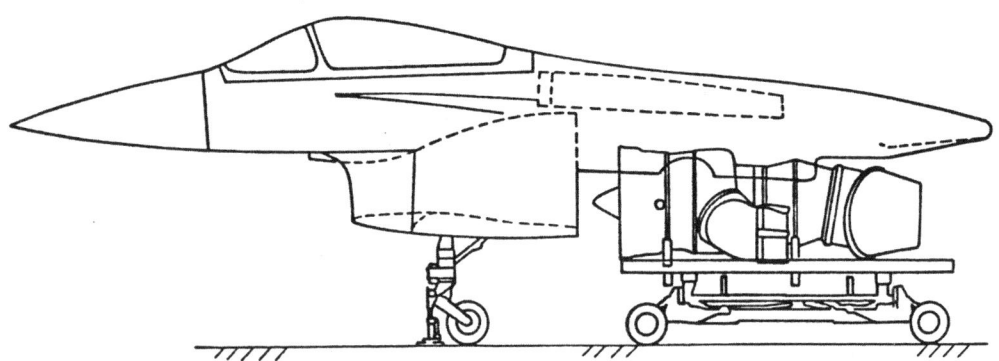

P.1216 engine-change schematic showing ease of access.

residing with Warton. The intention within MoD was that the 'winner' of Study B would become the next fighter for the RAF as well as equipping the Royal Navy, thereby achieving a requirement rather more cost effective than the small order for the Navy only.

The P.103 concept gathered few fans at the Directorate of Operational Requirements at the MoD. An assessment by an unnamed civil servant in DDOR/4 noted, 'Despite BAe's enthusiasm, I believe that the P.103 is a fundamentally unsound concept. If this really is BAe's perception of the only real alternative to the Pegasus concept then the only conclusion that I can draw is that if BAe are to produce a supersonic VSTOL aircraft between now and the end of the century, then it will be Pegasus based.' However, he wasn't that keen on P.1216 either. 'I find some aspects of the proposal disappointing. In particular the P.1216 … is difficult to trim in transition yet no mention is made of the potential offered by decoupling the front and rear nozzles as suggested by MacAir in their ideas for a demonstrator programme.' In conclusion he saw the P.103 'as a demonstrator for a concept that is not sound and the P.1216 I see as a rather shallow approach to an operational aircraft'. One wonders if the anonymous author had a sore head that day.[7]

The split of work across the two sites in 1985 did nothing to aid the closer co-operation between the northern and southern combines, various spats breaking out over essentially minor issues but highlighting the different work practices within management at Warton and Kingston. In considering whether a Harrier replacement needed to be supersonic, Ralph Hooper, in a memo to Mike Turner after his retirement, noted that the validity of the argument had become mired in a typical 'north/south' disagreement. 'Unfortunately, this question has become political via the Warton argument that Kingston can't be trusted to do a supersonic aircraft because they have never done one.' He might equally have pointed out that Warton had never 'done' a V/STOL aircraft.[8]

As the implications of the cost of a Sea Harrier replacement dawned on the Royal Navy, the lack of funding led to Sea Harrier FA.2 remaining in service for longer than originally envisaged, not leaving service until 2006, when the Harrier GR.7/9 fleet of the RAF under Joint Force Harrier was used as a replacement until its own retirement in 2011. Already, by the turn of the century it was becoming clearer that the job of P.1216 could be covered by land- and sea-based versions of the forthcoming Joint Strike Fighter F-35A and F-35B, developed principally in the US but with UK as a fully participating member. Ironically, the F-35B uses what might be termed a remote fan concept, once investigated by Kingston, though in the present case the front fan does not form part of the primary engine system, instead being a shaft-driven remote fan with its own intake system.

With regard to a replacement for Harrier GR.5/7 in RAF service, as seen above much work was carried out under NST.6464 leading in 1986 to an MoU between the UK, RAE and MoD, and the US, NASA and DoD, to explore ASTOVL concepts. The following year four studies were submitted covering earlier work – P.1230, replacing the twin-boom P.1216 with a single 'Harrier' type fuselage; P.116, ejector augmentation; P.112, RALS; and P.115, tandem fan. No design appeared to find favour over the others though the entire process raised Ralph Hooper's hackles. 'How the four competing designs are being assessed. In my experience it has always been extremely difficult to compare different projects on any basis that gave much confidence in the result … I find it hard to imagine a set-up less likely to produce a sensible decision than the one set up by BAe as you described.' By this time P.1216 had been dropped from the competition and replaced by P.1230, Hooper noting 'I am not clear if P.1216 is not being included in the study because it is too good or because it is too bad'.[9]

As Hooper reiterated when planning for a P.1216 brochure was in train, 'We should remind the reader that in the 26 years since work on jet V/STOL commenced at Kingston, we have examined every conceivable way of solving the problems of subsonic and supersonic powered lift fighter aircraft. If we are again back with a solution based on the "Harrier principle of vectored thrust" it is *not* because we have *not* investigated lift engines, inducer augmentation systems, flow switching systems, rotating pods or RALS but rather because we *have*', i.e., the various Warton-derived concepts.[10]

Ultimately the US lost interest in this particular path to ASTOVL and with no likelihood of US orders, interest in the UK also slipped away as Kingston Future Projects Office was closed and studies towards JSF came to increasingly be seen as the way forward.

If this aircraft was, as Kingston believed, a better mousetrap, why did the UK MoD and indeed the rest of the world not beat a path to its door? By the late 1980s international co-operation had become very much the name of the game. At Kingston, US collaboration had been the preferred route to any requirement for offshore participation while at Warton European collaboration had been foremost in recent projects with the Sepecat Jaguar and Panavia Tornado dominating their workload. With declining US interest in a

Prime Minister Margaret Thatcher inspecting the P.1216 mock-up at Kingston, 1982.

UK-derived ASTOVL project, and no real interest from Europe, Kingston was out on a limb with no suitable partner for P.1216 in the offing. At the same time, Warton's work on 'conventional' agile combat aircraft, the P.110, ultimately to lead, after many vicissitudes, to the EFA and Typhoon aircraft, was winning the battle within BAe for research and development funding; the key here was that as a multinational project, costs could be shared around the continent.

This question of cost would come to dominate MoD procurement. With an ever-decreasing slice of UK GDP to spend, the country's armed services had to consider very carefully the implications of their purchases. With P.1216, and its derivatives, lacking offshore collaboration, research and development costs and risks, not to mention any eventual purchase would fall squarely on the UK taxpayer, perceived value-for-money becoming the most important question to be considered.

On a somewhat more prosaic note, the internal politics of the BAe combine also had its effect. With the drift of influence within management flowing away from the 'Kingsbridge' enclave and towards the Warton-dominated stronghold in the north, Kingston's projects appeared to have seen their influence ebb away in the fight for resources that govern any large multi-site enterprise. This trend had its inevitable conclusion in, firstly, the closure of Kingston/Weybridge Future Projects Office and, very quickly, the entire closure of Weybridge and Kingston sites, followed by the closure of the remaining southern sites by 2000.

It is also possible that the unconventional layout of the P.1216 influenced some decision makers; it just didn't 'look' right. While this may seem a rather subjective conclusion, it should not be ignored. Certainly, it could be rationalised as concern regarding the wing-mounted booms with their aeroelasticity implications but, at bottom, it is still the case that military aircraft need to conform to human prejudices if they are to win orders. As the old saying goes, 'if it looks right, it will fly right'.

As a footnote to this story, perhaps the last word should be left to Chris Hansford, who, when told about a transfer to RAE Farnborough on the closure of the Future Projects Office in December 1988 commented, 'I have been trying to launch ASTOVL for over 25 years, and feel very sad that we don't have an ASTOVL EFA in prospect. Despite this background, changing threats and technology have caused me to re-examine my previous views; though still strongly favouring development of new-generation *subsonic* ASTOVL aircraft, I am no longer convinced of the case for *supersonic* ASTOVL in the post-EFA procurement cycle. For many of the major tasks (and particularly those few requiring supersonics), the economic and operational arguments for smaller, stealthier, unmanned vehicles appear incontrovertible, and the technology is imminently available', i.e., unmanned combat aircraft, dealt with later.[11]

Chapter 8

UFA: Unmanned Fighter Aircraft

The concept of controlling an aircraft remotely, without a pilot aboard, is almost as old as powered flight itself. Principally developed as military aids, their intended use ranged from aerial reconnaissance to attacks on enemy airships. Success with the primitive radio-control systems of the time was limited and it was not until the 1930s and, in particular, the advent of the Second World War that increased reliability allowed a more worthwhile use to be made. However, Sopwith Aviation had produced a very early radio-controlled aircraft, named the Sparrow, as an aerial target around 1915 but the design did not enter service, suggesting unreliability or a disinterested customer. In the UK, de Havilland developed a remotely-controlled aircraft in 1935 for use by the Royal Navy and Army as a gunnery target, based on the DH.82 Tiger Moth, and known as the Queen Bee. Later in the war, the USAAF developed a radio-control system for the B-17 Flying Fortress for use as a large flying bomb. Post-war, the Australian based Government Aircraft Factories developed the Jindivik drone, which first flew as a missile target in 1952, continuing in service for many years.

With the advent of improved remote-control systems, the drone returned to its earlier use as an unmanned reconnaissance asset, examples being used in the UK by both the Army and Royal Navy and, globally, by many nations. In the US one of the most ambitious developments was the Lockheed D-21 drone developed to be flown from the back of the SR-71 strategic reconnaissance aircraft as a means of extending reconnaissance over high-risk (militarily and politically) territory. Ultimately abandoned due to the poor results achieved, it nonetheless provided the US with a means to further the development of such aircraft.

With the development of the cruise missile in the US, achieved through constant refinement of inertial navigation systems, drones – UAVs or UCAVs – continued to be produced in ever more successful quantities. The control of such assets was now possible over extended distances, in the order of thousands of miles, due to the advent of satellite communications. The increasing sophistication of these aircraft would eventually allow them to be used not only

UFA: Unmanned Fighter Aircraft 173

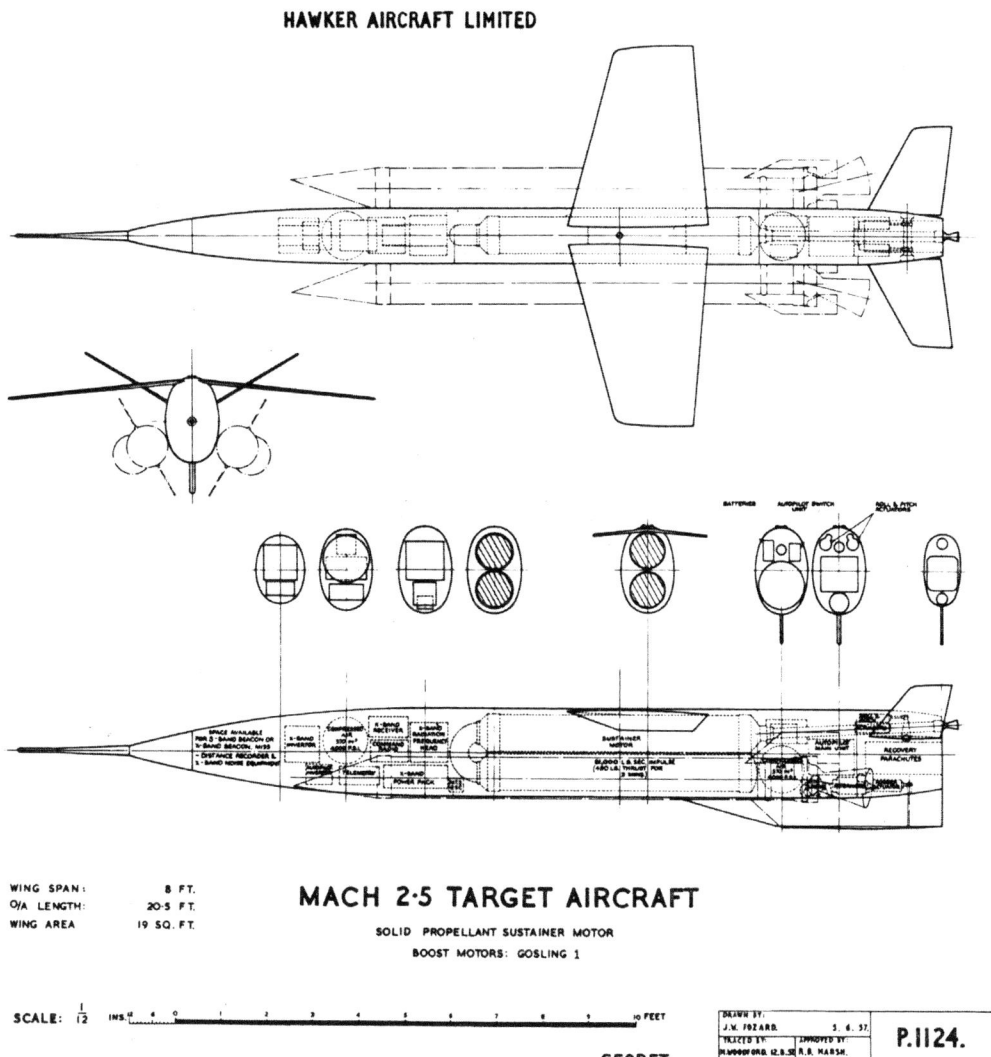

P.1124 GA, Mach 2.5 target aircraft.

for reconnaissance but as a weapon platform, seeing particular deployment by the US in the conflicts in Afghanistan and Iraq.

That Hawker and, later, Hawker Siddeley should become interested in such aircraft should therefore not come as a surprise. In May 1957 Hawker had allocated P.1124 to a Fozard design for a two-stage supersonic (Mach 2.5) rocket-powered target drone to be launched from a Hunter, to replace the Jindivik in operational service although this work had petered out by the end of the year. In 1962 a supersonic V/STOL drone had been mooted under the project code

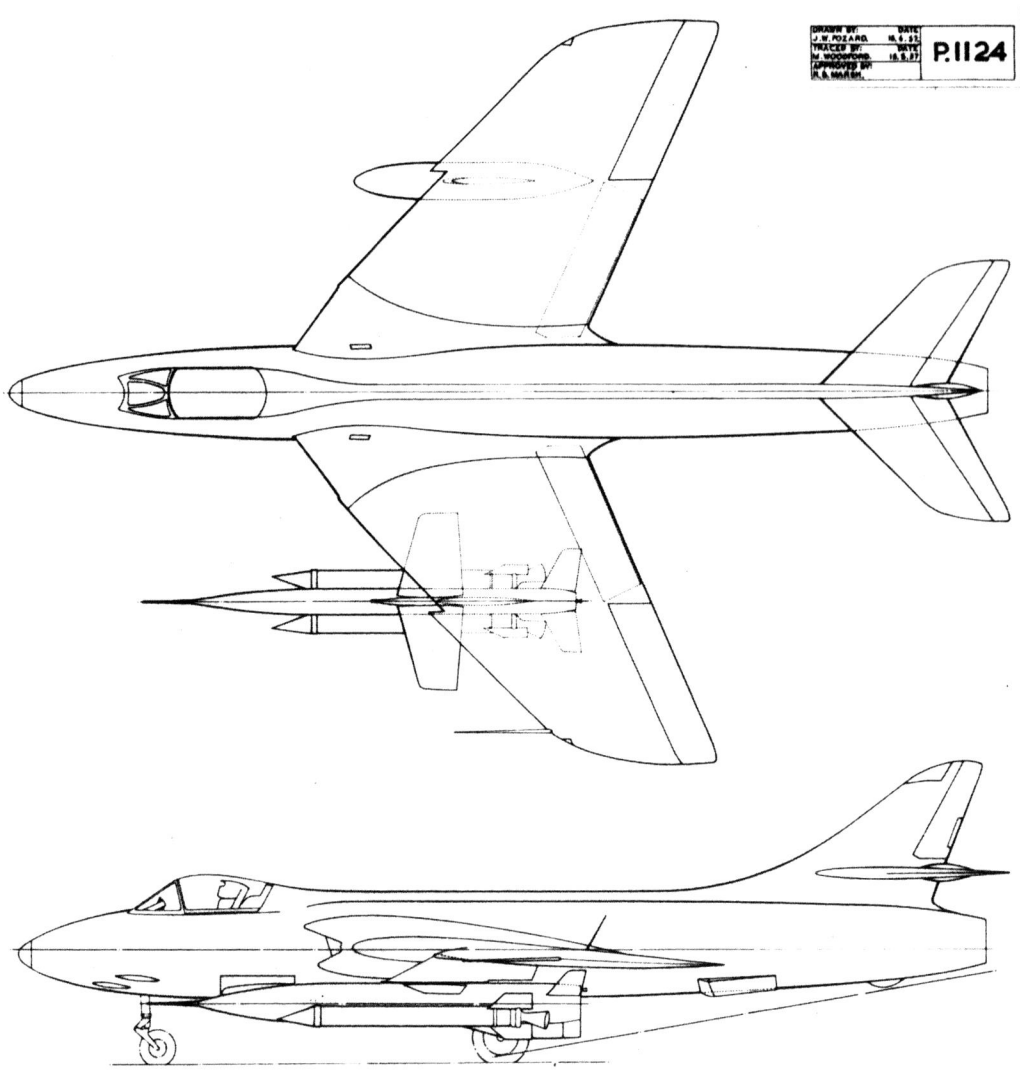

Hunter and P.1124 combination two-stage supersonic target system.

P.1157, powered by a Rolls-Royce RB.153 engine, but nothing further is known regarding this project. By 1973 Chris Hansford in Kingston's Future Projects Office, in an assessment of potential future projects, highlighted what he termed a 'drone airfield buster'.[1]

In a short resumé of the problem, he noted:

> It is clear that counter-air operations will be an important factor in any future European war. Airfield defences will be both numerous and sophisticated,

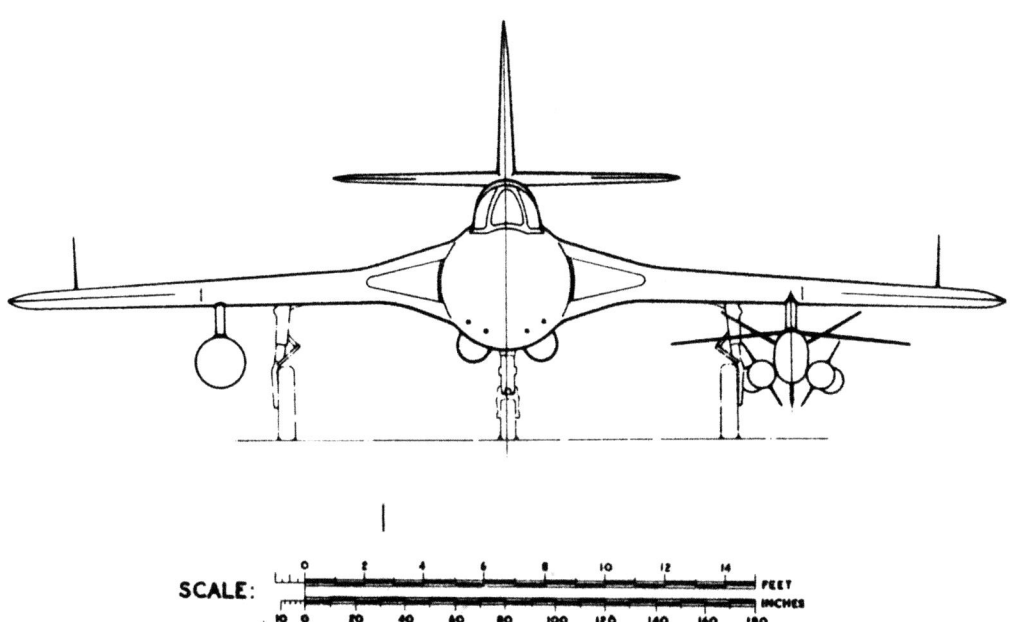

TWO-STAGE SUPERSONIC TARGET SYSTEM

HUNTER F.6 WITH ROLLS-ROYCE AVON TURBOJET
CARRYING P.1124 MACH 2·5 TARGET AIRCRAFT.

and hence even present estimates of loss rate for airfield denial missions fall generally in the 10%-15% range. If the enemy is assumed to develop a Rapier-like missile during the coming decade, 25% attrition might reasonably be estimated. With such high attrition rates envisaged, it is likely that this task could be fulfilled with greater cost-effectiveness by a much cheaper unmanned aircraft, leaving the manned aircraft free to attack the ground forces directly.

With the benefit of hindsight, it appears clear that what Hansford had in mind was what we would later understand as a cruise missile.[2]

Little work appears to have been carried out on the concept for the next ten years but in 1983 project number P.1224-1 was allocated for what was now termed the 'Ferret' drone or 'Unmanned Fighter Aircraft'. Chris Hansford's May 1983 appreciation of the concept began with the problem for which a solution might be engineered. 'Much is currently being made of the Warsaw Pact's improved capability in "look-down, shoot-down" fighters – it is claimed that the low-level

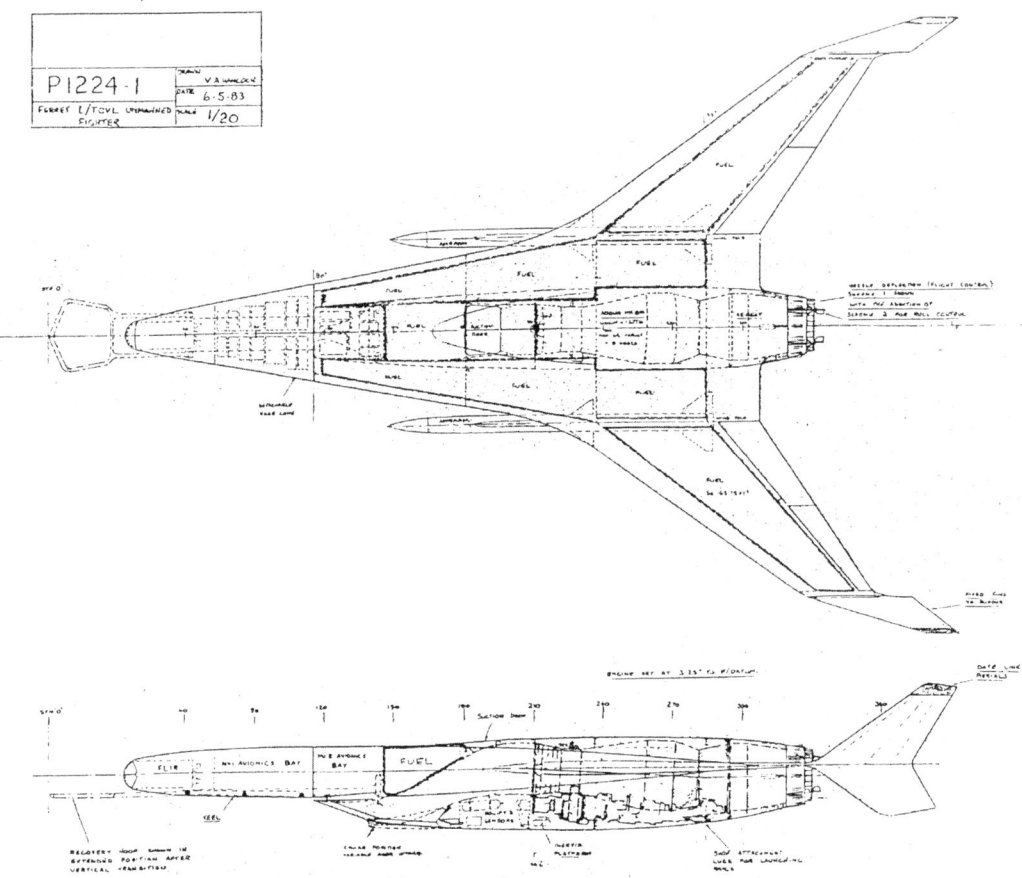

P.1224-1 GA, initial scheme for unmanned fighter, 1983.

attack option currently favoured by the RAF would no longer be viable in the face of such a threat. Leaving aside the extent to which this may or may not be true, contingency planning suggests that a counter to such fighter threats should be devised.' Hansford then went on to enumerate what appeared to be the most likely means of achieving this, including high-performance fighters, i.e., F-15, ECA, the 'missileer' concept, whereby a large subsonic aircraft armed with long-range missiles would outrange the enemy's missiles sufficiently to make the carrier aircraft's performance disadvantage unimportant, i.e., F-14 with Phoenix missiles; electronic countermeasures from either the intruding aircraft, AWACS or from the ground to disrupt the enemy missiles; possible satellite-based systems, e.g., lasers; anti-missile missiles carried on the intruding aircraft to counteract SAMs and MRAAMs; and, lastly, unmanned fighters 'programmed to shoot down *any* aircraft found within a particular area'.[3]

UFA: Unmanned Fighter Aircraft

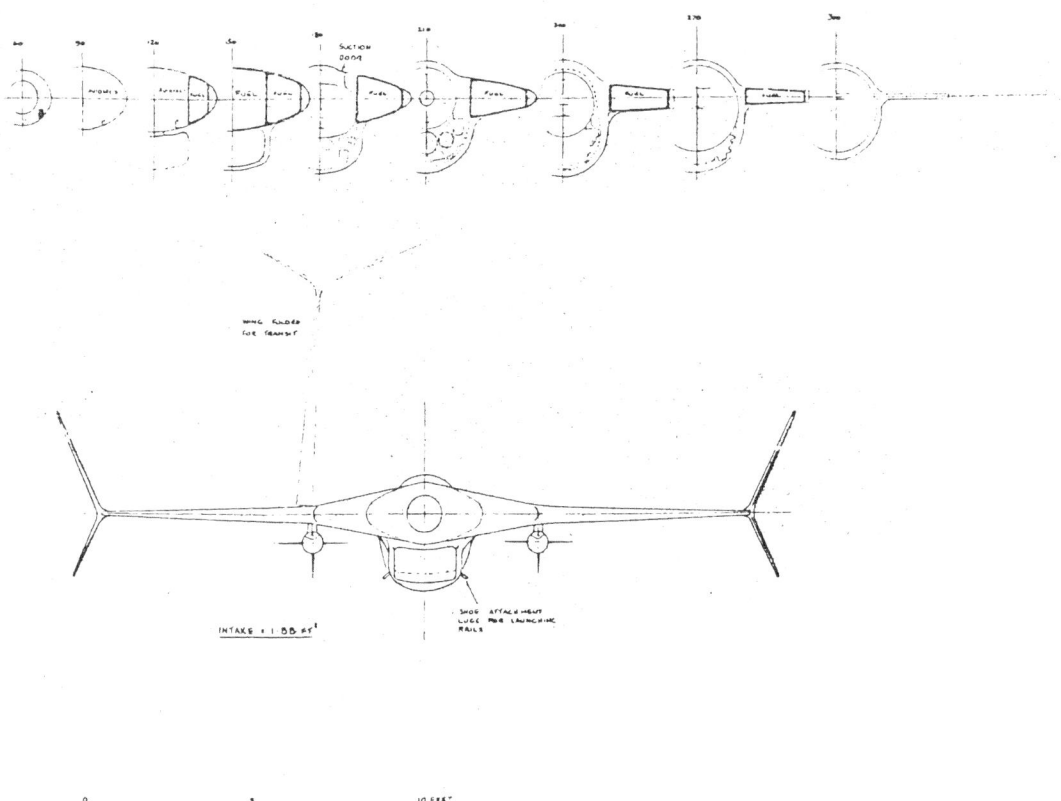

That the concern regarding the advent of Warsaw Pact fighters with a look-down, shoot-down capability was not just a BAe concern was revealed by Lockheed's Skunk Works boss Ben Rich who foresaw calamity awaiting the USAF General Dynamics F-111 bomber, its 'down-in-the-weeds' approach meaning that they would be sitting ducks as they flew into the battle zone at very low level, clearly visible to the Soviet fighters above.[4]

Vic Hancock's May 1983 appreciation showed a stealthy blended swept-wing aircraft of 15.3 feet span, wing area of 74.2 square feet and a length of 28.8 feet. Outward canted all-moving twin fins, described as elevons/flaperons, with similarly canted ventral fins offered the only moving control surfaces. However, engine nozzle deflection, mainly for the recovery phase and thrust vectoring in cruise might have been available for directional control. Initial thoughts on powerplant revolved around the Rolls-Royce Adour 811 reheated turbofan offering 8,400lb thrust fed via a chin-type intake. Armament was envisaged as missiles of unspecified type carried under the wing. In keeping with the

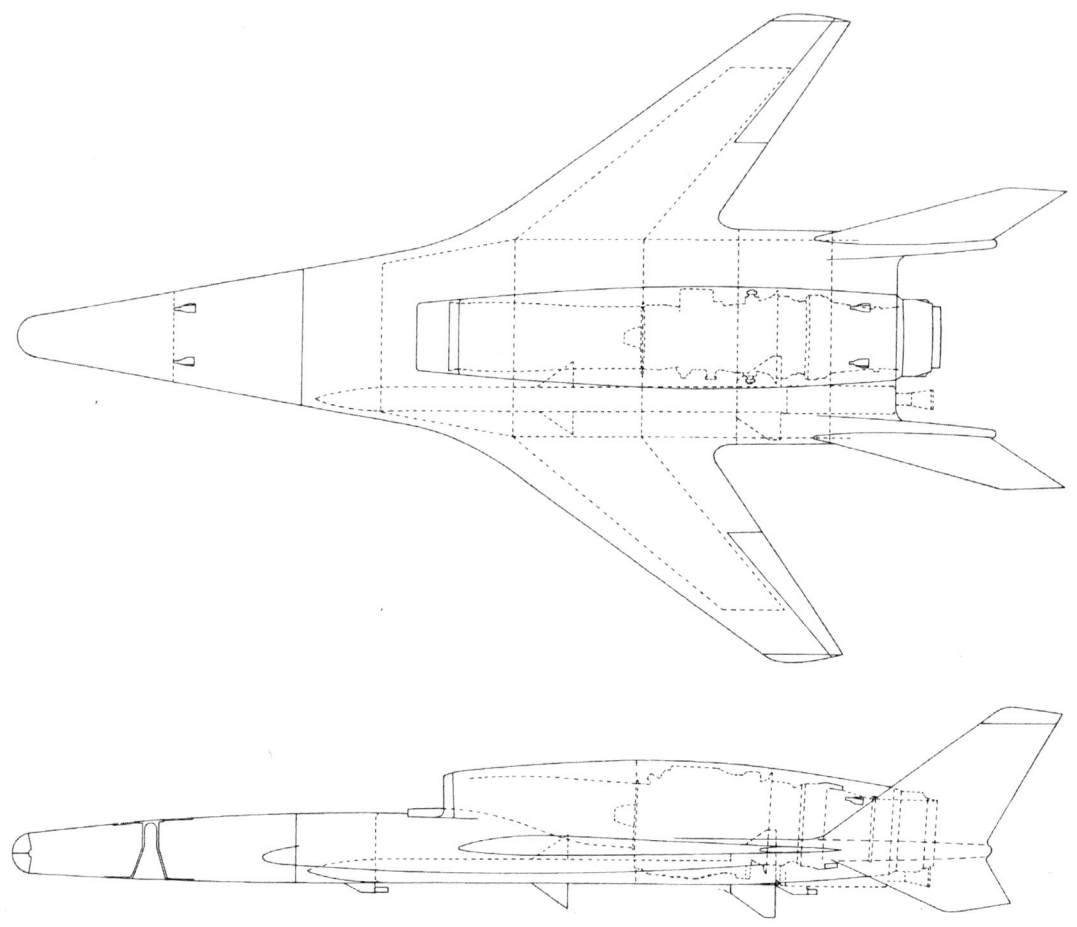

P.1224-2 GA, subsonic unmanned fighter powered by Viper Mk.680 turbojet.

requirement for weight saving, no undercarriage was intended, the aircraft being zero launched from a mobile transporter using additional Blackcap rocket boosters offering 14,850lb thrust. However, what might seem to be an overly complex recovery procedure was envisaged whereby the aircraft would assume a vertical aspect at low level, made possible by the swivelling nozzle directing the jet efflux, and capture by the mobile transporter via an extendible loop in the nose to pick up the transporter gantry. On a successful capture, the aircraft would be lowered for servicing and re-use. It appears that the recovery phase would probably have been by far the most exacting of the entire sortie.[5]

With the aircraft mission by now seen as an unmanned fighter rather than ground attack weapon, an early amendment to this layout saw the main and

1224-2	Subsonic Unmanned Fighter Aircraft (UFA)

Engine Viper Mk 680-43

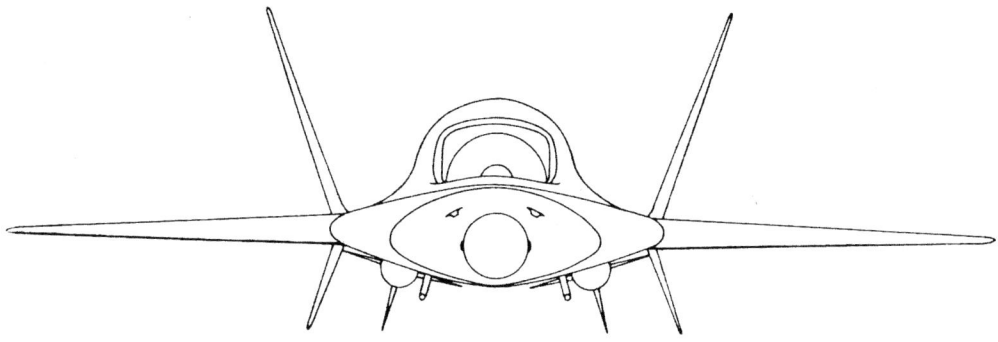

ventral fins moved out to the ends of the wings which had increased span of 19 feet. Missile armament was now specified as AMRAAM; wing slats and ailerons were added in addition to the all-moving fins to increase controllability and the swivelling exhaust annulus was confirmed as being capable of (limited) yaw and pitch control, mainly for the recovery phase where it would be the only effective control. Sortie performance was predicated on an empty weight of 5,705lb and an all-up weight of 8,645lb, a range of 150–250 miles with a climb to 45,000 feet and an acceleration to Mach 1.4, loiter time not specified, and assumed sufficient fuel remaining for a recovery.

In considering what type of transporter might be required, an articulated cab and trailer were envisaged with sufficient launch rail/track carried to allow either an 83-feet rocket boosted or a 135-feet unboosted launch at 5-degree inclination. For a recovery, the trailer-mounted rail would assume a vertical position surmounted by a hook-on mechanism onto which the vertical aspect aircraft would latch for successful recovery, much in the way that US companies attempted in the 1950s, in particular Ryan with the X-13 Vertijet VTOL research aircraft.[6]

In July 1983, in a fuller report on the concept and initial thinking within Future Projects, T. Woods and Vic Hancock noted the changed requirement at which the aircraft was now targeted. 'The unmanned fighter aircraft is designed to penetrate enemy territory and intercept look-down, shoot-down aircraft ahead of Ground Attack aircraft. The unmanned aircraft is small and of low radar cross-section and so will sense enemy radar primary emissions before its own reflection is strong enough to be detected. The P.1224 is programmed to search a specified area or target illumination can be via AWACS. Secondary use in the intercept, ground attack and reconnaissance roles may enhance manned aircraft operations.' Sensors were intended to be passive (combined FLIR/Laser-ranging and passive radar sensor, i.e., ESM) to enhance the stealth characteristics, although the absence of an active radar posed questions about the mid-term guidance phase of the preferred AMRAAM missile. In considering the preferred sensor fit, Hansford's May 1983 note accepted that 'the lack of target information is currently the weak point of the concept', to which the recipient of the note had responded 'you can say that AGAIN'. Hansford in concluding his note with a summary of the characteristics stated, 'alternative suggestions are required'. To this the recipient had pencilled 'FORGET IT!' And, in case his acerbic rejoinders had been missed, added 'Experience shows that *total* unmanned complex is inflexible and *more* expensive than manned solutions.'[7]

In T. Woods' analysis of September 1983, a subsonic version of the aircraft, P.1224-2, had reverted to twin fins carried on the rear fuselage rather than

P.1224 UFA in flight, artist's impression.

wing tips, the Adour engine replaced with a Viper turbojet fed by an over-wing intake, and twin AMRAAM carried semi-submerged under the wing/fuselage joint. The lower performance was offset by the aircraft being smaller and lighter, and an over-wing engine intake offering a smaller radar cross-section. Launch attitude had now increased to 60 degrees, necessitating rocket booster assistance for all launches. A version of the -2 was envisaged as a small AEW platform, the basic aircraft having provision for larger bolt-on wings and carrying a 100-inch radar scanner on its back. In the P.1224-3 of October 1983 the engine reverted to the Adour with reheat and retained the over-wing intake while the -5 version offered the above plus active radar sensors. In 1984 P.1224-4 Hancock envisaged a larger AEW aircraft based on the -3 planform but with a much-increased aspect ratio, the wingspan being 52 feet and fitted with conventional flaps and ailerons and outboard spoilers, while the length had increased to 28.75 feet, the fins were fixed rather than moving and a rudimentary undercarriage was fitted. Engine selection was suggested as the Rolls-Royce RB.401-06. Carried above the fuselage was a 100-inch radar dish to perform the AEW requirement.

An overview of 1986/87 sought to summarise the requirements for a successful UFA as follows:

Critical Technology Areas
1) Air Vehicle Configuration. Stealth is vital – particularly low radar signature.
2) Sensors. Passive sensing of enemy emissions required, desirably 1° accuracy.
3) Communications. Narrow bandwidth secure comms link UFA – UFA highly desirable. (Also desired for MFA – UFA and ground – UFA links).
4) Computing. Large data storage requirements, at low cost and volume. Algorithms for target discrimination and prioritization.
5) Weapon System. Missile inertial mode, autonomous target detection without update. HOJ improvements and/or hand-off capability desirable.
6) Controllability. Integrated control system, using in-flight nozzle deflection. Roll control using servo driven by engine HP bleed. Quick and level transition to vertical attitude.
7) Recovery. Guidance during final approach – passive system preferred. Modified Skyhook technique for final capture.[8]

By October 1987 the configuration of UFA had changed radically. P.1224-7, produced after the Aerodynamics Department investigation of the -6 configuration revealed changes required, was now drawn with an ogival low-mounted wing planform with fixed twin fins at the rear of the fuselage to provide the optimum low RCS. Control was achieved via continuous-blown ailerons which should remain effective even in the hover with engine nozzle vectoring of up to 17 degrees axial deflection being available. Length was now 26 feet 2 inches and span a mere 14 feet and area 145.83 square feet. The powerplant remained the Viper turbojet and the missile armament had sensibly been relocated to the upper wing area to further reduce RCS.

In a preliminary specification reflecting current research on the UFA concept in July 1988, Hansford reiterated that the intention was to stick with passive sensor technology, using this to detect and target enemy radar emissions, and hence targets, prior to the UFA itself being 'seen' by hostile radar due to its small size and stealthy configuration, thereby achieving a missile launch before the opposition could respond. In listing its advantages, he identified:

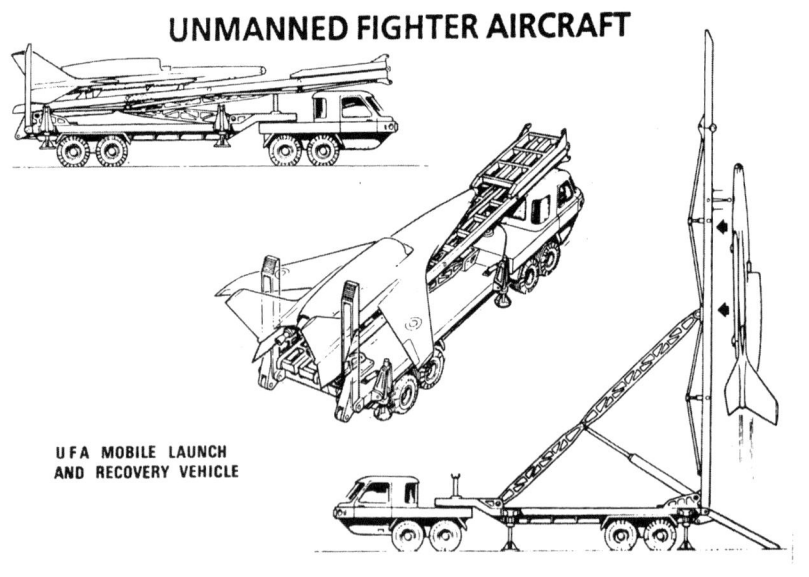

P.1224 imagery illustrating potential launch and recovery sequence.

a) Dispersed operations
b) Automated launch, engagement and recovery
c) High effectiveness
d) Low initial cost
e) Very low life cycle costs
f) Very high cost-effectiveness

In summary, UFA is a 'low-cost' system, designed to achieve an undetected first shot capability in Beyond Visual Range (BVR) combat.[9]

By this time, 1988, the aircraft planform for P.1224-8 had been 'rounded off' and reduced in size in a further nod to stealth. The -8 was now 17 feet long with a span of 11 feet and wing area of 75 square feet. The twin fins appear to have been deleted while the vectoring nozzle's movement was increased to 24 degrees. The last iteration of UFA under P.1224 appears to be P.1224-10 which follows the layout of the -8 in all but detail. The last design for UFA was P.1243 of September 1988 which broadly followed the stealth attributes applied to P.1224-10. But it would appear that the concept was taken no further, certainly not at Kingston.

In assessing the efficiency of the UFA concept, there were certain similarities between this and Lockheed's F-117A Nighthawk. This aircraft also relied entirely on passive sensor technology to complete its mission, principally laser guidance and FLIR. However, despite its popular moniker, 'Stealth Fighter', the aircraft was actually used in the light bomber role to destroy high-value assets in pinpoint-accuracy attacks. Although transit to and from the target was highly automated, the targeting was completed manually by the pilot. What Kingston was proposing was a leap from ground attack to counter-air missions, both offensive and defensive, a wholly more complex requirement. Now, the scenario was for UFA to fly sorties to dislodge Warsaw Pact fighters from their CAP positions 100 kilometres behind the FEBA to open the way for conventional air assets to transit to their targets or as an escort to ground-attack aircraft. To achieve this, UFA would penetrate the FEBA at high altitude and air speed in order to evade some at least of the inevitable SAM threat in the forward battlefield area. The concern was that the stealthy UFA might still be visible to the CAP Flankers before UFA was in a position to attack. Alternatively, UFA could fly CAP missions on its own side of the FEBA to counter approaching enemy fighters.

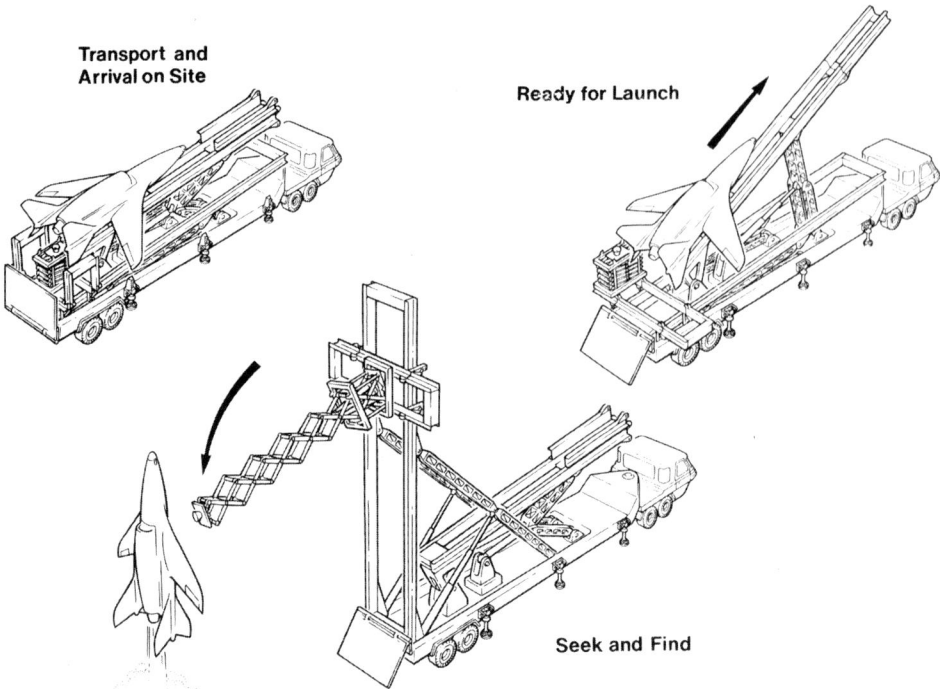

P.1224 imagery illustrating potential launch and recovery sequence.

In an assessment of 1988, by BAe Warton, in assuming that the principal threat would be from agile BVR-missile-equipped fighters, the two major concerns highlighted were:

1) The combat effectiveness of the aircraft against the anticipated Warsaw Pact air superiority fighter threat.
2) The viability of the passive sensors in the extremely dense Central European signal environment.[10]

Perhaps a third concern would be the control of the asset in combat. While autonomous systems could be used for transit to and from the combat area, active control by a ground-based pilot would require very secure data links, immune to jamming or manipulation. Therefore, the avionics element of the UFA fell into various discrete sub-sets.

1) Those required to allow manual control of the aircraft without interference from the opposition (or indeed from unintended allied interference). It will be apparent that the secure control of unmanned

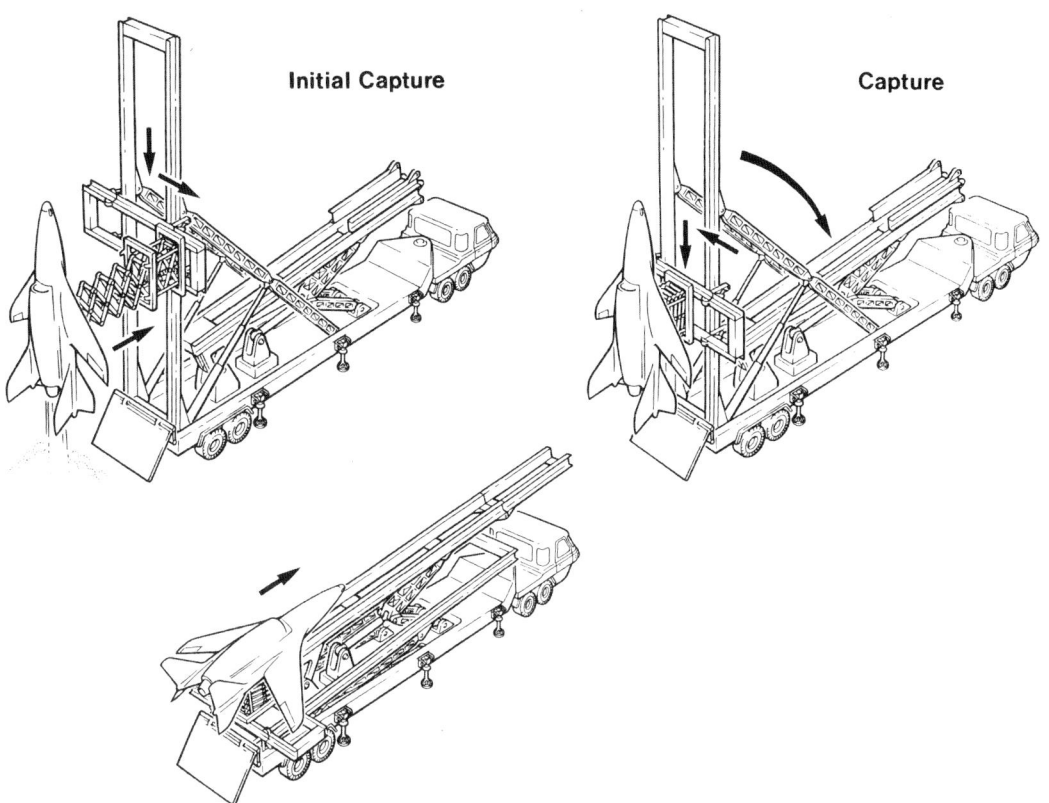

P.1224 imagery illustrating potential launch and recovery sequence.

assets is now achievable with present day drones, though the degree of signal security available in the late 1980s is a moot point. It has not been possible to determine just what degree of autonomous control and just how much, if any, interaction with a human pilot was envisaged by the design team.

2) Those required to allow the acquisition of targets. The ability of the UFA to acquire its targets was predicated on emissive emanations from that target (i.e., radar, jamming or radio signal traffic) and/or from infra-red (IR) emissions caused by engine exhaust or airframe heating. If the opposing forces were flying with their radars off and no radio traffic, UFA would clearly struggle to acquire targets at longer ranges, though in a battle situation, this scenario would seem unlikely. Warsaw Pact (WP) pilots would need active radar to acquire their own targets and were heavily controlled by ground-based commands.

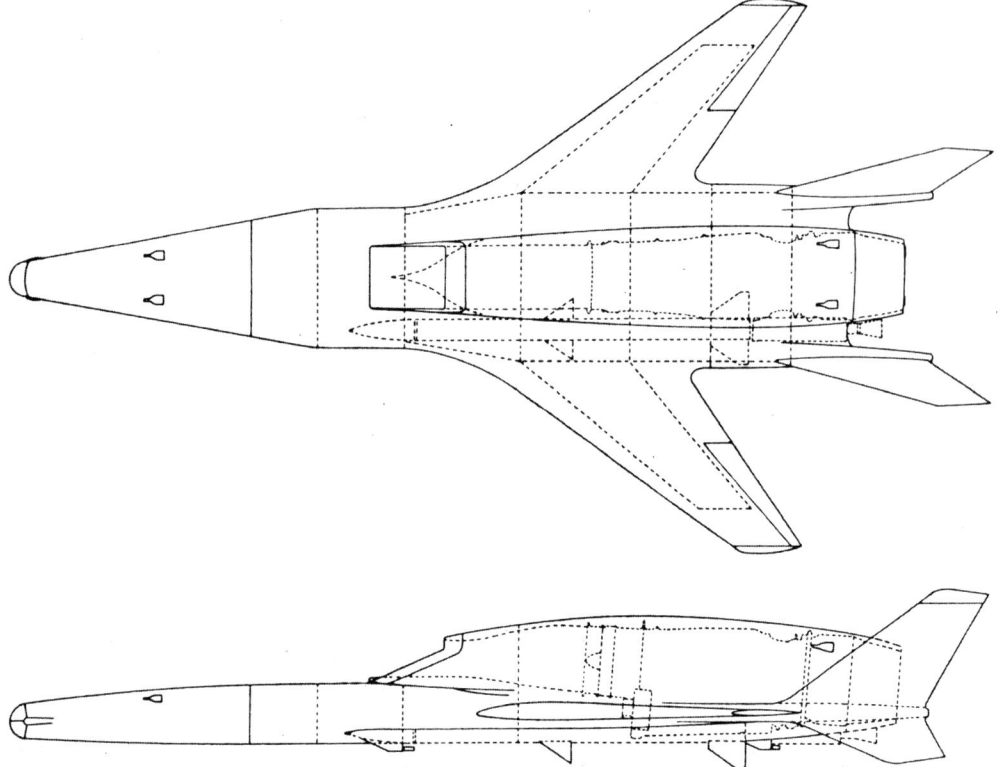

P.1224-3 GA supersonic unmanned fighter powered by RR Adour 811.

3) Those required to survive and win in combat situations. UFA's armament comprised AMRAAM-type BVR missiles. Normally these would be under the control of the parent aircraft's radar for mid-term guidance updates, only becoming autonomous in the terminal phase of an attack. Here perhaps lies the weakness of UFA as a 'fighter'. The aircraft avionics would need the ability to identify a target, acquire it, track it and accurately guide a missile to it to achieve a kill. Against a manoeuvring target, the missile would require mid-term command updates if it was to have any chance of successful terminal phase guidance and therefore a 'kill'.

In any event, further assessment determined that there were no technology 'stoppers' and that a flying demonstrator would allow many of the concerns to be investigated and resolved. In February 1988, a two-part demonstrator programme was proposed. Phase 1 would 'demonstrate successful launch from

P1224-3	Supersonic Unmanned Fighter Aircraft (UFA)

Engine – Adour 811 with reheat

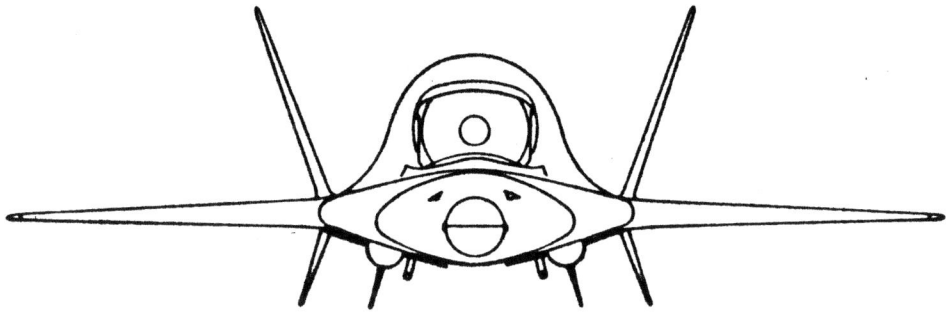

a ground vehicle, in-flight control, and return to and capture by the Skyhook-based recovery system.' Phase 2 would 'demonstrate the detection, acquisition and practice attack of radar-emitting target aircraft, using passive target sensors.' The demonstrator was never built.[11]

The sensors required to make UFA an effective weapon were perhaps the most complex research aspect involved with the concept. To achieve target detection without giving its position away meant a reliance on passive technology. This limited the choices to RWR/EMS and IRST. Use of a passive radar-warning receiver with electronic-support-measures ability would allow the detection of hostile aircraft radars and provide coarse tracking of the target in any weather. Infra-red search and track coupled to laser range-finding would again allow detection and tracking but would be limited by weather conditions, cloud cover over the central region was a distinct possibility and would severely limit laser ranging. It might also alert hostile aircraft equipped with laser-warning receiver (LWR) equipment.

Assessment of the likely AI radar in use by the Warsaw Pact aircraft (Slotback) suggested that even a stealthy UFA would be detectable at around 25 kilometres and at a similar range using IRST. Therefore, UFA had to be capable of achieving a detection and missile launch at greater than this distance, 30 kilometres being considered possible. Using a medium-range missile such as AMRAAM, this range was considered suitable provided that the probability of

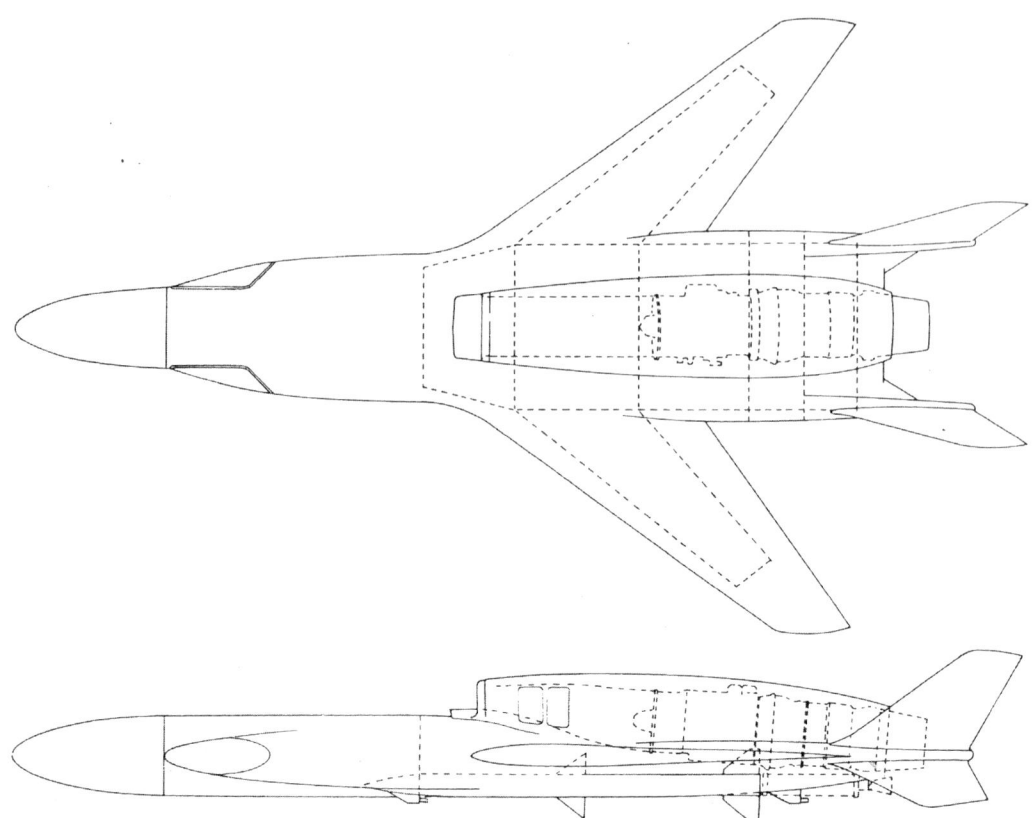

P.1224-5 subsonic unmanned fighter with active radar fit.

kill was acceptably high. The problem with a UFA/AMRAAM combination was that the most accurate method for successful interception was 'command/inertial/active' allowing mid-term guidance for the missile; this would not be possible due to the requirement for an active data link which would be detectable. However, AMRAAM also had an 'inertial/active' mode whereby the missile could be guided by inertial navigation to within short range, when active terminal guidance would complete the actual interception. Clearly, the terminal guidance phase had to be kept as short as possible because it would be detected by the target's RWR at this point.

After the autumn of 1988 UFA disappears from the radar. This was another project that appears to have met its demise with the closure of Kingston Future Projects. Henceforth, Future Project leadership would, by default, be based at Warton which, at the time, was becoming committed to EFA, the European Fighter Aircraft, that eventually reached service in the RAF as the Typhoon. The concept was perhaps ahead of its time. However, Warton has continued

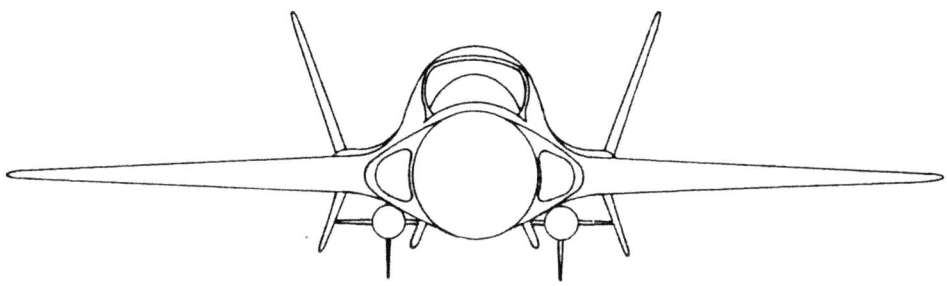

with UMA research and has built at least one technology demonstrator, though no aircraft, to the author's knowledge, has entered service with the ability to carry out the UFA mission. It is possible that Tempest, a project being shared by BAE Systems, Leonado UK, MBDA UK and Rolls-Royce, will go some way to producing a UFA-type aircraft. This project, currently in the project design and development phase, offers an unmanned future combat air system for entry into service around 2030. Whether such an expensive project can survive the inevitable cost cutting endemic in the UK defence procurement world is a moot point.

Although UAVs now form a comprehensive part of the world's air arms, these tend to be mainly designed for surveillance rather than offensive operations. While US drones have acquired an offensive ground-attack capability with smart missiles, this is a very different use to the counter-air role envisaged by the Kingston team for their UFA. Ultimately, it is likely that a combination of lack of an official requirement and of sufficiently intelligent avionics, coupled with an unquantifiable financial return on investment, made UFA a step too far for the aerospace corporation that BAe had become.

Perhaps the last word should go to Ralph Hooper. After his retirement, he said of UFA, 'I feel no warmth I'm afraid. But then, I got into aviation because I always knew I would enjoy flying – and I can't see why anyone would want to go to war if they weren't going to get some fun out of it. I mean, rape and pillage isn't encouraged anymore and if they cut out the flying too, you might as well stay at home and help with the black market! … If Kingston goes this way, would it ever get back into real aircraft I wonder? What about Dunsfold? If it can succeed – well, good luck to it.'[12]

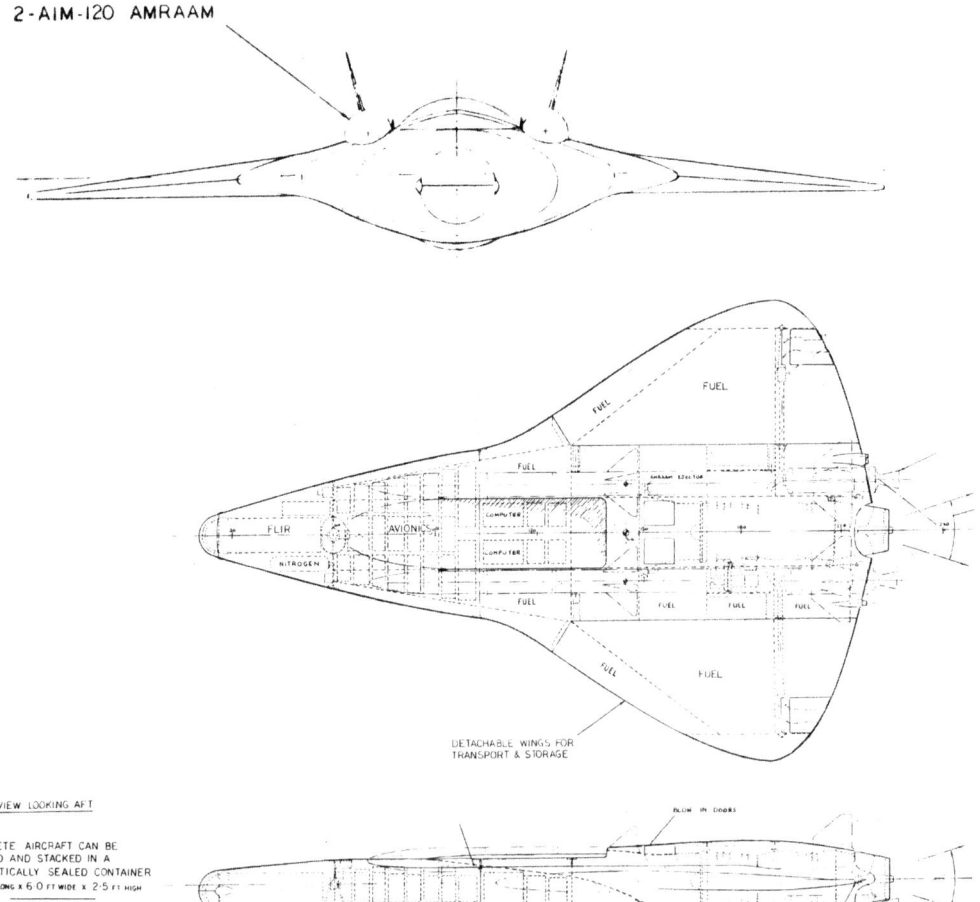

P.1224-8 GA, unmanned fighter final configuration.

Chapter 9

Conclusion

At the end of 1988 Future Projects at Kingston ceased to exist as an independent entity; Weybridge had ceased to exist the year before. Two years later, the complete closure of the Kingston site was announced and, in 1999, that of Dunsfold, completed in 2000. Thus ended Surrey's aviation manufacturing connection that had endured for ninety-two years. In that time over 45,000 Hawker/HSA/BAe Kingston aircraft had been designed and manufactured, both at Kingston/Dunsfold and at many other aircraft factories, excluding AV-8B/GR.5 and T-45; indeed, over 11,000 were built by other concerns.

In terms of new projects coming forward and entering manufacture, the numbers show a steady decline over the period of Hawker's tenure at Kingston, reflecting the continually increasing complexity and cost of the end product. Although the Project Office staff was deliberately kept to small numbers throughout the period, the other design functions saw a steady increase in numbers to cope with the increasing workload. As seen earlier, in the 1920s, design staff numbered around 31; by the start of the Second World War, this had increased to 146. By the mid-1980s this number had increased to over 800. Over the same period, new Hawker designs achieving flight declined from seventeen in the period 1920–31 to three in the period 1961–70, two in 1981–90 and none in the 1990s, excluding new marks of existing aircraft.

This was not due to any failure on the part of the company but rather reflected the incredible growth of complexity of the military aeroplane. The entire procurement process had also seen this same complexity leading to extension, from maybe two years from initial design to service entry in the 1920s to today's extended process which can stretch out up to ten to fifteen years and see products entering service which are already effectively obsolete. Systems, and avionics in particular, accounted for much of this delay. In many cases basic research was required to produce new avionic systems before any development phase could begin. In the 1960s, in particular, electronic progress was rapid with the changeover from analogue to digital technology and the miniaturisation that could then flow from this, again leading to the requirement for extended research and development of new products. The somewhat cumbersome ministry laboratories that were responsible for much of this research could be, at best, ponderous in their work output to such an extent that commercial developments of similar products could have been substituted and thereby shortened in-service dates had the Air Staff and MoD been amenable to this. But the most extensive growth of all has been in the paperwork that now accompanies any attempt to design and launch a new aircraft project, and the bureaucracy that has inevitably grown to administer this particular mountain.

Another change that has been evident over the period has been the change of emphasis in terms of the generation of new requirements. In a simpler age, it was quite common for an aircraft manufacturer to approach the military authorities with a new design which, if it was considered of merit, would have an OR written around it. Now this is hardly ever the case and design requirements are generated by smart graduates in the MoD, often on short term tours of duty from the RAF or RN, and disseminated to industry in a strictly one-way street.

With regard to Kingston's design function, it could be said that there was a definite divide, one might say into two different time periods, this divide marked by Camm's death. Sydney Camm had dominated the company for so many years that at times he must have seemed immortal and, as the Chief Designer of the UK's most successful aircraft company, it seemed that he dominated the entire industry. That this domination declined with advancing years was perhaps inevitable but, within the company and without, his presence was always felt, whether he was leaning on a designer's board telling him where he had gone wrong or drinking tea with an air vice marshal in Whitehall, telling him much the same thing. There was no love lost among the civil servants in the MoS and MoD for Camm, more than one relishing the idea of Hawkers getting a bloody nose over some project. The fact that Camm's design for the Hurricane was probably responsible for their being free to pursue their lives was no doubt lost on them. But, because Camm was such a titan, his death inevitably left a vacuum which was never really filled. Kingston's Design Office lost its single overriding (overbearing?) focus urging them on to greater things, never accepting second best, while the company lost its irascible champion, always pushing the company point of view with the movers and shakers in government and the services. It was Camm's personality that often carried the day as much as the merit of the design.

His death occurred in the time when management practices were changing, US ideas of running a company coming to the fore, encouraged by the relevant ministries with which Hawker had to deal. There was less room for the giant personalities who had driven the industry previously and anyway, who could replace Camm? In the immediate hiatus, this fell to Roy Chaplin and, on his retirement, to Ralph Hooper and John Fozard, both extremely able and talented designers but their management styles were very different to what had gone before. Hooper's social diffidence perhaps made him less able to step into the big man's shoes than Fozard, who appeared to be indestructible, and both were given executive roles in which they were very successful, but the chance to mould another Camm was perhaps deliberately avoided by the HSA board who had also felt the lash of Camm's tongue on more than one occasion.

The projects that have been examined here all failed for one reason or another. All were good designs that could have successfully fulfilled the role envisaged for them during the Cold War. Why then did they fail to reach production? This was not a specifically Hawker experience; all companies, in the defence business suffered similarly, often for reasons outside their control.

The period encompassed by the Cold War – the late 1940s through to the late 1980s – was one in which the political, financial and social climate of the UK changed beyond recognition. The Second World War had depleted the country financially to such an extent that it would remain in hock to the USA for years to come. This penury was reflected in what the governments of the day believed they could achieve with what slender resources were available to them. Sometimes they got it right; often they got it wrong. Financially, from 1945 to the 1980s, the Government was forced to balance its income by allocating finance across the various ministries where it was perceived to be most needed. The slice that came to the Ministry of Supply/Ministry of Defence was then channelled into whichever projects were considered to be militarily and politically the most urgent. At times, this balancing act was upset by either a particularly expensive project, i.e., development of atomic weapons and creation of the V Force, by political expediency – cancellation of military projects such as TSR.2 to release funding for social projects – or market forces, the financial crisis of 1967 which saw the devaluation of Sterling against the US Dollar. The slender finance available to the Defence Ministry resulted in each aircraft project being weighed to ensure that, in the minds of the Air Staff and the Treasury, it was the most cost-effective means of providing the function it was designed to fulfil.

By the 1960s it was widely accepted that there were too many aircraft companies chasing too many projects, few of which would reach fruition. This situation ultimately resolved itself into the consolidation of the industry into five large amalgams of aircraft and engine manufacturers, of which Hawker Siddeley was probably the largest. Fewer, larger projects became the order of the day although the finance arrangements of the government/industry relationship took longer to resolve with the move from 'cost-plus' (government covering the cost of production plus a fixed company profit percentage) to 'fixed-price' contracts where the company was responsible for cost overruns.

Politically, the period was exemplified by the dismantling of the British Empire. Socially this was widely recognised as the right course of action, although the trick would be to retreat from the Empire with the minimum of bloodshed. In most cases this was achieved successfully but in India with the creation of Pakistan, religious differences resulted in widespread loss of life. Allied to the social requirement for retreat was the burden upon the country of military support 'East of Suez'; the eventual withdrawal from this requirement allowed the military budget to be reduced and redirected to the European theatre. This change in emphasis affected directly the aircraft projects being designed and procured: no longer would a global stance be needed; the military situation in northern Europe could be the main focus of interest but, in terms of aircraft sales, this would be a reduced focus. After 1945 the UK aircraft companies had a seller's market available as European and newly independent countries sought to rebuild or acquire air forces from the redundant stocks of wartime airframes and newly developed jets. This situation did not last; both the US and a resurgent French industry began to make inroads into the former UK markets, further squeezing company profits and the concomitant export credits.

In its attempts to define just what the future military threat, and hence the military response should be, the UK armed forces faced yet another upheaval, this time in technological progress. The post-war period witnessed a phenomenal growth and evolution in advanced technology to such an extent that there was a real danger of bringing aircraft, in particular, into service that were already obsolete. The result of this realisation was wholesale cancellation of projects as they failed to live up to the early promise of the technology. While the Labour administration that cancelled TSR-2, P.1154 and HS.681 in 1965 has taken much ire over the years, the previous Conservative government had in fact been responsible for far more project cancellations. The only answer to this appeared to be to try to jump ahead with designs, to achieve a 'quantum leap' that would leapfrog a generation of design, in the hope that what eventually entered service was still of some use.

What this quantum leap might look like meant different things to different people. With the realisation that the defensive systems of the USSR would decimate any fleets of high-flying nuclear bombers with surface-to-air missiles (SAMs), attention increasingly turned to the use of missiles, both to deliver nuclear stores and to engage enemy aircraft over friendly territory. The arguments vis-a-vis manned aircraft versus guided missiles for air defence came increasingly to the fore in the 1950s, culminating in the political decision in 1957 to re-align the UK's defences away from manned interceptors towards surface-to-air guided missiles. Unfortunately for the country, this decision was promulgated by Duncan Sandys whose wartime experience had convinced him that the future belonged to guided missiles; there would no longer be a requirement for manned aircraft with which to defend the UK. This decision would come to colour the thinking of the Air Staff for the next decade and result in many poor decisions that would reflect on the future procurement of military aircraft, culminating in TSR-2. This promising project eventually came to carry so much political and military 'baggage' that its ultimate failure doomed it to become the most raked over decision of the past fifty years.

So much for the national background, but what of Hawker's experiences? It is interesting to examine the projects featured here and note the changes across the nearly forty years represented in terms of power, speed, weight, armament, role and, of course, cost.

In terms of available power, the Hunter, designed in the late 1940s, had emerged at the dawn of the axial-jet turbine and began life with the Rolls-Royce Avon RA.7 and AS Sapphire Mk.101 of some 7,500lb thrust in 1953. Later this was upgraded to the 10,000lb-plus series 200 engine. Hawker's next product took the company into the realms of far more powerful engines in the race to achieve a supersonic interceptor. The projected DH Gyron offered for the P.1103 and P.1121 doubled the promised thrust to 20,000lb but ultimately failed along with Hawker's aeroplane. With the advent of Hawker's V/STOL investigations, the engine that became the Pegasus grew from around 10,000 to 21,500lb for the Harrier, this being limited by the ability to increase the fan dimensions, constricted as it was by the Harrier fuselage size. Finally, the Rolls-Royce RB.422-60 was scheduled to produce something in the region of 40,000lb thrust for the P.1216 which would have provided the aircraft with a very spirited performance. In the early days of the jet engine, the engine manufacturers were always 'behind the curve' in their ability to provide sufficient power at acceptable weights and specific fuel consumption but, by the later decades of the century, this problem was largely solved and excess power became the norm. Over this same period, fuel consumption saw a steady decline with the increased efficiency of available engines.

Aircraft speed over the same period saw a similar evolutionary, rather than revolutionary, increase which was partly a function of available engine power and partly that of aerodynamic design. As was seen with the Hunter development, its potential speed was governed more by the limits of the wing design than by available engine performance, limiting it to transonic speeds in level flight. It was as much the inability to fit sufficient fuel into the existing Hunter airframe to power the reheated Avon that doomed the further development of a missile-armed interceptor, rather than any aerodynamic limitations. The P.1121 was designed as a Mach 2 interceptor and, had the Gyron/Hawker intake combination worked satisfactorily, and sooner, it is quite possible that the Air Staff might have been persuaded of the aircraft's potential. Certainly, based on the case of the Republic F-105 Thunderchief, a US contemporary of the P.1121, the aircraft would have found a suitable service role for many years. P.1154 was also offered as a Mach 2 aircraft although, if it had entered service, this would more likely have been Mach 1.7. In terms of speed, the aircraft was entirely dependent on BSEL mastering successfully the intricacies of PCB (the Pegasus version of reheat) and there were those at Bristol who were quietly relieved when the project was cancelled, such was the uncertainty surrounding the concept. P.1216 was again conceived as a Mach 2 fighter and, by this time, the requirements of a PCB system were better understood, not least due to extensive testing of both models and full-size aircraft in the Shoeburyness test rig. It will be seen then that Mach 2 was viewed as a natural limit to fighter aircraft design over many years. Largely, this was governed by the thermodynamic qualities of aircraft-grade aluminium alloys which degraded quickly at speeds over approximately Mach 2.2. It was for this reason that Concorde's optimum cruising speed was kept at Mach 2.02. To achieve greater speeds, the aircraft manufacturer needed to conquer the special requirements of working in metals such as stainless steel and titanium.

Aircraft weight over the Cold War period was certainly always regarded as a problem, to be set against available engine power. For CTOL aircraft increasing weight was seen as inevitable and an irritant to be fought against. But, for VSTOL aircraft it could be the difference between successful flight and failure to get airborne at all. The Hunter was built with generous allowances in frame and skin thickness resulting in an extremely resilient airframe; in the VSTOL designs such latitude was simply not available, even up to the P.1216 where

Hooper was aghast at how the aircraft had put on weight, a penalty that was just not acceptable if it was to win the ASTOVL battle.

In terms of aircraft role, those projects reviewed here were pretty consistent across the five decades represented. Hawker Aircraft Ltd had, throughout its existence, concentrated on fighters with a strong secondary role of light bomber/ground attack. The Hunter was conceived as a pure fighter but, by 1960, was really obsolete in this role. Fortunately, Hawker had, from the earliest days, planned in the aircraft as a ground-attack machine and, in this guise, the FGA.9 Hunter went on for many years as a highly successful mount in multiple air arms. P.1103 had been offered as a fighter, or interceptor as it was by then termed, but, when this failed to find favour with the Air Ministry, was recast as the P.1121, a high-speed low-level strike weapon. P.1154 was regarded mainly as a strike platform with a secondary interceptor role, while P.1216 continued this theme, and reversed it for the naval version.

With the arrival of the Hunter, weapons packages were still predominantly based on massed guns, though the rifle-calibre weapons of the Second World War were now obsolete, being replaced first by 20mm and then by 30mm cannon. The four-30mm cannon pack of the Hunter formed a formidable striking force against an unwary adversary but suffered the same sighting problems as those of the war years, namely the ability of the pilot to bring the firepower to bear on the target using only an optical sight in the split second that might offer itself at the high speeds involved. The answer to this was seen as a guided missile and research was underway in the 1950s to achieve such a weapon with sufficient reliability to give the possessor a realistic advantage over the opponent. Early trials with a radar-beam-riding missile showed that, while successful interception was theoretically possible, the realities of successful deployment were less so. This weapon fit came with cumbersome electronics and a radar pack, both prone to overheating. The infra-red homing missile, however, offered a more realistic weapon for successful interception and was to prove fairly simple to develop and has formed a basic component of air interdiction since the 1960s. Both of these missile fits were trialled on Hawker's Hunter but never achieved service entry with the UK. From this time, guided missiles were seen as the weapon of choice for air defence and, also at times, for the strike role, though only in more recent years has the air-to-surface guided missile found greater favour against the dumb bomb. By the 1980s missile combat in the air had reverted to radar-guided missiles, though with the passive infra-red missile as the favoured close-in weapon system. For the UK, long-range air-to-air missiles (LRAAM), in the guise particularly of the Hughes AIM-120 AMRAAM, were seen as the weapon of choice for BVR engagements, with the Warsaw Pact having similar weapons available. Thus, it was now possible for opposing interceptors to join battle in the skies over the battlefield, having never seen the opposition.

It is of interest to note that, as in former days when the Spanish, the French or the Dutch were considered as *the* enemy over many decades, so, in the second half of the twentieth century, it was the USSR – Soviet Russia and its satellites – which was accepted as the most imminent threat to the wellbeing of the UK and NATO. The physical manifestation that this threat represented evolved over time and Hawker's projects evolved similarly in an effort to counter that threat.

During the late 1940s/and early 1950s, the threat was seen to be long-range piston-engined bombers, Tu-4s, Soviet copies of the Boeing B-29 Superfortress, approaching the UK's east coast in massed waves to drop conventional and, later, nuclear bombs on cities and centres of industry. This was the threat that the Hunter was designed to defeat. To counter this, aircraft were specified capable of meeting that threat quickly, directly as it was detected, over the North Sea, before it could make landfall. Assuming that the USSR had no fighter capable of long-range escort missions, it was hoped that these fighters would fall upon the approaching air fleet and destroy it with massed cannon fire before it could begin its attack.

For this reason, a long-range requirement for UK fighters was not considered a priority and, consequently, 1950s fighters suffered from small fuel capacities which limited endurance. Once the Soviet Union was in possession of nuclear weapons, it became manifestly obvious that the hoped-for enemy attrition rate of 10 per cent was almost pointless; it would be necessary to destroy 90 per cent of the incoming threat to stand any chance of avoiding total annihilation of UK infrastructure; a missile-armed supersonic Hunter in this scenario would certainly have improved the chances of obtaining a higher rate of attrition. The decision to decline the missile-armed P.1983 Hunter development was a serious error in terms of military defence but, at the time, the pursuit of a more powerful Hunter with the 'big' engine was seen as a cheaper and more reliable alternative to the immediate need.

By the late 1950s P.1121 was offered as a formidable supersonic fighter and ground-attack platform, a next generation Hunter, armed with radar-controlled missiles for collision-course interceptions using Red Hebe, a very large air-to-air missile or by development of the DH Blue Jay IV, later named Blue Vesta, entering service as Red Top. In the ground-attack role, tactical nuclear or conventional bombs could be delivered in high-speed dashes to pummel the attacking forces. With the advent of the 1960s, nuclear warfare, both strategic and tactical, from the first few days of conflict was considered to be inevitable. The failure of the P.1103/P.1121/P.1129 project may be seen as the result of striving for 'the best' when 'good' will do. The failure of the Air Staff to acquire *most* of what it wanted, in pursuit of the illusion that it could have *all* it wanted in the shape of TSR.2, ultimately resulted in the failure to acquire either. This situation was mirrored by the Royal Navy, the dream of a fleet of large carriers based on CVA-01, perceived as threatened by acquisition of P.1154, was similarly dashed with the summary cancellation of the entire carrier project. Of course, the ultimate sanction for all these military projects was that wielded by the UK Treasury, a body singularly ill-equipped to assess the use or otherwise of military projects.

The major threat to NATO was seen to be massed attacks of tanks and infantry across the German border on the European Central Front. P.1154 was designed to counter this threat by operating in the tactical low-level role, to deliver a nuclear weapon in the rear of approaching enemy forces. The ability to deliver these weapons from dispersed locations without a requirement for airfields, which would have been destroyed in the first days of war, was seen as a potent weapon by both sides. The continuation of the project after the exit of the Royal Navy offered the RAF a flexible weapon system tailored to the NATO environment. Its cancellation removed its potential from the order of battle with no adequate replacement. The acquisition of Phantoms was a useful replacement in the air defence field but, as a strike weapon, it was never going to replace the capabilities of P.1154.

The Vietnam War perhaps showed how the helicopter had ceased to be a sideshow to military conflict and was now a central part of battle planning. The Soviet Union, in particular, took notice of the opportunities that attack helicopters offered and developed a formidable array of craft capable of swift low-level penetration which conventional aircraft struggled to counter due to the helicopter's high manoeuvrability and nap-of-the-earth flight capability. Aircraft such as the Mi-24 Hind were recognised as a problem requiring an urgent solution and this is what SABA was intended to offer. In time, however, it was the ground-to-air missile such as the US Stinger that came to be regarded as the helicopter's main opponent. Carried by infantry in the field, these lightweight cheap missiles effectively neutered the advantage of the attack helicopter to a significant degree.

UFA was considered as an antidote to losses of air assets in high-intensity conflict situations. Losses of valuable air crew were now considered as being as unacceptable as losses of hugely expensive aircraft. What was needed was a weapon that avoided both pitfalls. UFA was pilotless and designed to be as cheap as possible. The threat in the 1980s was seen particularly as that provided by Soviet fighters armed with new radars able to scan below from high level, pick out aerial targets from the ground clutter and then shoot them down with missiles able to do the same. This look-down-shoot-down technology was something that NATO considered to offer a potent threat to allied aircraft; UFA could counter this by being sufficiently stealthy to see the enemy before the enemy saw it and use its own intelligent missiles to neutralise the threat. While the unmanned fighter, as opposed to reconnaissance drones, did not see conflict in Europe due to the temporary ending of the Cold War, its abilities are still under development and may yet be seen in future battle zones.

In summary then, the question must be asked: why did the projects considered here fail to reach production? Recurrent causes emerge, lack of funding, lack of political will, lack of cohesive service agreement, or interest, and constantly changing defence policy largely accounted for these failures. Added to these was the failure to get to grips with the constantly changing technological base upon which these projects were founded. In essence then, this work is perhaps best described as a study in failure.

With regard to costs, it is interesting that costs at Hawker and at HSA Kingston/Dunsfold had always been tightly controlled, Camm, in particular, considering parsimony as a useful string to management's bow. At times this could verge on the ridiculous. Of all the large aircraft companies, Hawker/HSA never possessed a wind tunnel, Camm considering it a needless extravagance. He much preferred to cadge time in the wind tunnels at RAE Farnborough or other HSA companies which, although ostensibly 'free' meant that Hawker work had to join the back of the queue for access. Similarly, when mainframe computers became more common, a new computer suite was constructed at Dunsfold, not in a purpose-built facility but squeezed into an old wartime Nissen hut hidden in the trees. It was not until BAe was privatised that funding, agreed during its existence as a nationalised concern, eventually began to flow into Dunsfold sufficient to finally move the facilities into new buildings and demolish the wartime Nissen hut village that had been their former home.

The designs considered here are all post Second World War, due partly to the author's bias, but also to the fact that these projects were far more complex and costly and, therefore, more important for the company's existence compared to pre-war designs. As such, their failure to reach production became a disproportionate burden upon the company and, some might add, the country. The efforts of Hawker Aircraft and Hawker Siddeley Aviation Design Office as they became in 1960 broadly fell into discrete areas characterised by the needs of the RAF on the one hand and those of the company on the other. These areas may be summed up as follows:

1) developments of the basic Hunter,
2) creation of a Mach 2 fighter,
3) development of a supersonic VSTOL fighter, including lift-engine technology.

While other projects were certainly proposed by the PO, these would be the areas where most design effort would be expended, ultimately unsuccessfully.

Finally, the cancelled projects need to be seen in the context of Hawker's successful post-war projects. The profitable production runs of the Sea Hawk, Hunter, Harrier and Hawk certainly combined to keep the factories humming for fifty years and firmly cemented the place of Hawker Aircraft and its successors in the story of aircraft production in the twentieth century.

Appendix

Hawker, Hawker Siddeley Aviation & BAe Kingston Project Numbers 1940–1988

Project Number	Description	Date of Issue
P.1000	Shaft Drive Single Seat, single engine	
P.1001	Shaft Drive Twin Engine Convoy Fighter	
P.1002	20mm cannon installation (Hurricane IIc)	
P.1003	Henley Escort Fighter (Boulton Paul turret)	
P.1004	High Altitude Fighter F.4/40. To AM.4/40	
P.1005	High Speed Bomber (Sabre x 2). To AM.12/40, 3/41, 7/41, 9/41	
P.1006	Henley Close Support Bomber. B.20/40. To AM.12/41	
P.1007	High Altitude Fighter (single seat). Not submitted	
P.1008	Night Fighter F.18/40. Not submitted	
P.1009	Typhoon Fleet Fighter conversion (Sabre) to N.11/40. To AM.5/41	
P.1010	Typhoon Turbo Blower. To AM.6/41	
P.1011	P.1005 with Power Jet engine added. To AM.6/41	
P.1012	Typhoon II. To AM.9/41	
P.1013	Remote guns in P.1005	Sept 1941
P.1014	Power Jets Engined Fighter. Not submitted	
P.1015	P.1005 (Centaurus)	February 1942
P.1016	Typhoon II (Griffon)	April 1942
P.1017	Griffon Fighter with T.2 wings	
P.1018	Light Fighter (Sabre) (Fury). To AM.9/42	
P.1019	Single Seat Light Fighter (Griffon with contra-prop). To AM.9/42	
P.1020	Light Fighter (Centaurus) (Fury Mk.1). To AM.9/42	
P.1021	Tempest Development (Centaurus). To AM.11/42	June 1943

Project Number	Description	Date of Issue
P.1022	Single Seat Fleet Fighter (Centaurus) (Sea Fury Mk.X). F.2/43 to N7/43. To DD/RDT/5/43	January 1944
P.1023	Sabre 44 in Tempest I	January 1943
P.1024	Tempest Development (Sabre)	
P.1025	Griffon Light Fighter. To DTD	September 1943
P.1026	Griffon in F.2/43. To DTD & AD/RDT	
P.1027	R-R.46-H-42 in Tempest (contra-prop)	July 1943
P.1028	Tail-less Fighter	August 1943
P.1029	Tail First Aeroplane	
P.1030	4,000hp Fighter (R-R.46-H-42) (contra-prop)	September 1943
P.1031	R-R B.40 Jet Fighter	
P.1032	R-R 46.H. F.2	April 1944
P.1033	2 x R-R 46H Bomber (P.1005)	
P.1034	2 x R-R B.41 Bomber (P.1005)	July 1944
P.1035	1 x RB.41 in F.2/43	
P.1036	Sabre V in F.2/43	April 1944
P.1037	2 Griffons in Twin boom	September 1944
P.1038	B.41 in version of P.1034	September 1944
P.1039	2 x B.41 in body	October 1944
P.1040	R-R B.41 bifurcated pipes. To MoS.2/45	December 1944
P.1041	Mosquito Replacement (General Development)	January 1945
P.1042	Variation of P.1040	July 1945
P.1043	No undercarriage - P.1040	April 1945
P.1044	FAA Fighter Bomber	June 1945
P.1045	Single R-R AJ.65 Interceptor	July 1945
P.1046	FAA version of P.1040 with rockets	October 1945
P.1047	Extreme Sweep Back on P.1040 with rocket	August 1945
P.1048	Twin AJ.65 Interceptor (straight wing) (Tendered)	February 1945
P.1049	Single R-R AJ.65 Interceptor with extreme Sweep Back	January 1946
P.1050	Long Range, High Altitude, Tail-less Transport	February 1946
P.1051	Fleet medium bomber, 2 x AJ.65 (7,000 lb missile)	
P.1052	P.1040 with sweep back to E.38/46 (R-R Nene). [Flew as sweep back research aircraft]	
P.1053	Rocket Fighter	

Project Number	Description	Date of Issue
P.1054	Interceptor Fighter. 2 x AJ.65s (close to body). Sweep back on wings and delta tail. To F.43/46	
P.1055	Eight-seat Commercial Light Transport. Armstrong Siddeley AS. Mamba x 2	
P.1056	Land Night Fighter (OR.227). 2 x R-R Avon. Straight wing. To F.44/46	December 1946
P.1057	Land Night Fighter (OR.227). 2 x R-R Avon. Swept wing, delta tail. To F.44/46	December 1946
P.1058	Light 4/5 Seat Private Owner/Taxi	January 1947
P.1059	Naval Night Fighter, Straight wing, outboard nacelles	April 1947
P.1060	Naval Night Fighter, Straight wing, recessed nacelles	
P.1061	P.1054 with straight wings. 2 x R-R Avon	May 1947
P.1062	Interim Interceptor Fighter, P.1052 with AJ.65. Swept wing, delta tail.	October 1947
P.1063	N.9/47	October 1947
P.1064	Two Avons on top of wing	January 1948
P.1065	Fighter with one engine and 2,000lb rocket	February 1948
P.1066	[missing]	
P.1067	F.3/48 specification. R-R AJ.65 Avon. [Flew as Hunter Prototype]	April 1948
P.1068	Interim Fighter. P.1040 with R-R Nene and straight pipe, plus reheat	September 1948
P.1069	Transonic Fighter. R-R Avon or AS Sapphire plus reheat	September 1948
P.1070	Transonic Fighter	October 1948
P.1071	Transonic Fighter. (R-R Avon plus reheat and 2,000lb rocket)	October 1948
P.1072	P.1040 with AS Snarler rocket. [Flew]	December 1948
P.1073	"Monsoon" Interceptor Fighter. R-R Tay. (Was P.1072 then altered)	January 1949
P.1074	N.7/46	February 1949
P.1075	P.1062 with Nene II or R-R Tay plus reheat	February 1949
P.1076	Investigation based on P.1067	May 1949
P.1077	General Purpose Tailless Fighter, two Avons	July 1949
P.1078	P.1052 with AS Screamer rocket	August 1949
P.1079	Not used	

Project Number	Description	Date of Issue
P.1080	Australian P.1052	April 1949
P.1081	Australian Interceptor Fighter P.1062 with R-R Nene. [Flew]	
P.1082	Fighter to F.23/48?	May 1950
P.1083	50% sweepback on P.1067 [Supersonic Hunter development. R-R RA.14 + reheat]	June 1950
P.1084	Delta Fighter used for investigations. Single engine, large fin	April 1951
P.1085	Delta Fighter used for investigations. Single engine, large fin	April 1951
P.1086	Not used	
P.1087	Naval Interceptor version of P.1081	April 1951
P.1088	Light Fighter 2 x 3,000lb thrust engines	August 1951
P.1089	Rocket Fighter. 5,000lb thrust. AS Screamer or DH Spectre. To F.124.T	August 1951
P.1090	P.1083 with DH Gyron	August 1951
P.1091	P.1067 with delta wing. AS Sapphire 4 plus reheat	October 1951
P.1092	Supersonic all-weather fighter (blended Delta) R-R Avon	November 1951
P.1093	Single seat supersonic all-weather fighter (Delta) single seat. R-R Ra.14 or DH Gyron	February 1952
P.1094	P.1072 with thin wings. R-R Nene and AS Snarler rocket	March 1952
P.1095	P.1083 with larger fuselage to take Sa.4 or RA.14 & 2,000k	June 1952
P.1096	Supersonic Research Aircraft. Delta with tail RB.106. To ER.134.T	May 1953
P.1097	Supersonic Research Aircraft, P.1083 Development. RB.106. To ER.134.T	May 1953
P.1098	Medium Transport, 2 Alvis Leonides	July 1953
P.1099	Hunter development with RA.19, 23 or 28. Hunter F.6 prototype. [Flew]	August 1953
P.1100	Supersonic Hunter. RA.24 + 2 x 2,000lb rockets and wing tip tanks	April 1953
P.1101	Hunter trainer. (T.Mk.7). R-R RA.23 engine	July 1953
P.1102	Thin Winged Hunter Development	October 1953
P.1103	Single engined, two seat Mach 2 Interceptor (DH Gyron)	January 1954
P.1104	Two engined Mach 2 Delta (and semi-delta) Fighter (R-R RB.112)	January 1954

Project Number	Description	Date of Issue
P.1105	Hunter F.6 with podded Napier rocket boost	March 1954
P.1106	Thin winged, larger span missile armed Hunter with AS Sapphire Sa.10	May 1954
P.1107	Straight thin-winged Mach 2 fighter with 4 or 6 x 21in dia engines	July 1954
P.1108	Naval Strike Fighter to M.148T. 4 x RB.115 engines	September 1954
P.1109	Hunter F.6 with Blue Jay and AI.20	
P.1110	Omitted	
P.1111	Omitted	
P.1112	Omitted	
P.1113	Omitted	
P.1114	Two-seat all-weather Hunter development with wing tip tanks. Avon engine	November 1955
P.1115	Two-seat all-weather Hunter development with wing tip tanks. Sapphire engine	November 1955
P.1116	Mach 2 Interceptor and Long Range Strike Fighter. DH Gyron engine	May 1956
P.1117	Naval Hunter with AI.23 and Blue Jay missiles. R-R Avon RA.24 engine	March 1956
P.1118	High Speed Hunter with straight wing and tail. R-R Avon plus reheat	March 1956
P.1119	High Speed development P.1103	March 1956
P.1120	Two-Seat Advanced Hunter Trainer. R-R Avon Mk.122 engine	July 1956
P.1121	Air Superiority Strike Fighter, single/twin seat, based on P.1103 (Mach 2 plus)	May 1956
P.1122	Steel wing version of P.1121	January 1957
P.1123	Mach 2 Tactical Bomber, two Seat version of P.1121	January 1957
P.1124	Two-Stage Supersonic (Mach 2.5) Target system	April 1957
P.1125	Twin-Engine (R-R RB.133 Medway) Supersonic Strike Aircraft, Development of P.1121	June 1957
P.1126	Subsonic VTOL Strike Aircraft, cranked delta wing. (R-R lift engines + two Orpheus)	June 1957
P.1127	Subsonic V/STOL Strike Aircraft (Bristol BE.53 engine). [Flew 1960]	July 1957
P.1128	Hunter Jet Transport. 5-Seat Pressurised version of Hunter (Two BE Orpheus Turbojets)	October 1957

Project Number	Description	Date of Issue
P.1129	2-Seat Supersonic Strike Aircraft to GOR.339 (2 x R-R RB.141R)	November 1957
P.1130	Indian All-Weather Fighter based on 2 seat Hunter	December 1957
P.1131	Long Range Large Freighter (4 x R-R Tynes)	March 1958
P.1132	V/STOL Strike Aircraft (2x Bristol BE.53 plus Napier Double Scorpion rocket) or twin boom version minus rocket	April 1958
P.1133	Hunter F.6 with AI.23 and Firestreak. R-R Avon Mk.203	August 1958
P.1134	Research Aircraft for Mach 3-4. (R-R RB.146 + reheat plus 2 x ramjets)	December 1958
P.1135	Thin Wing Hunter with R-R RB.146	January 1959
P.1136	Canard VTO Strike Fighter (subsonic). 4 x R-R RB.153 lift engines plus 1 x cruise	April 1959
P.1137	Tilt nacelle Supersonic V/STOL Tactical Aircraft. 2 x RB.153 lift engines in nacelles plus 3 x engines in forward fuselage plus 2 x clang box lift/cruise engines	July 1959
P.1138	Mach 3 VTO canard Naval Fighter with 6 x R-R RB.153 lift and 3 x RB.153R cruise engines, or Mach 3 Military Aircraft with ogival wing and 4 x RB.153R engines	September 1959
P.1139	Subsonic VTO Strike Fighter (2 x R-R RB.153 lift engines and RB.163R cruise engine with clang box). 2/60	February 1960
P.1140	Supersonic VTO Strike Fighter (3 x R-R RB.153 lift engines and 1 x RB.163 cruise engine with clang box and reheat)	March 1960
P.1141	Supersonic Strike Fighter. Bristol BE.53/11R engine plus VG variant	May 1960
P.1142	Supersonic STOL Aircraft. Bristol BE.53/1	
P.1143	Supersonic V/STOL Aircraft with reverse delta wing. 6 x R-R RB.153 lift engines and 4 x RB.153 in wing tip pods.	July 1960
P.1144	Supersonic V/STOL Strike Aircraft. 2 x RB.163 with reheat & 6 x RB.162	August 1960
P.1145	Supersonic STO Research/Strike. 1 x DH aft fan Gyron Junior	July 1960
P.1146	Supersonic V/STOL Strike Aircraft. 2 x RB.173 with reheat plus 6 x RB.162. (Development of P.1144)	December 1960
P.1147	STO Naval Strike/Interceptor. (2 x R-R RB.173 plus 4 x RB.162)	December 1960
P.1148	STO Naval Strike/Interceptor with variable sweep. (Engines as P.1147)	January 1961

Project Number	Description	Date of Issue
P.1149	Supersonic V/STOL Strike Aircraft with variable sweep. 2 x R-R RB.168R lift/cruise VT plus 6 x RB.162 lift engines	February 1961
P.1150	Supersonic V/STOL Development of P.1127 with BS.53 engine (3 stage fan plus PCB), Pegasus 5 or BS.100 engine	January 1961
P.1151	Supersonic OR.346 2 seat Aircraft with jet flap. 4 x R-R RB.153 engines plus reheat	April 1961
P.1152	Supersonic RAF/Naval V/STOL Strike Aircraft. R-R RB.177 plus reheat cruise and 8 x RB.162 lift engines	June 1961
P.1153	Supersonic V/STOL Strike Aircraft (OR.346). One Olympus 22 with reheat	June 1961
P.1154	Supersonic V/STOL Strike/Recce Aircraft to NBMR.3. RAF/RN options with single or two seats. BSE BS.100/9 with PCB or Pegasus 5 with PCB or BS.100/8 with reheat. [Cancelled 1965]	October 1961
P.1155	Supersonic V/STOL Strike/Recce Aircraft based on P.1150 and P.1154. Bristol Pegasus 5 with PCB plus 2 x RB.162	October 1961
P.1156	Subsonic V/STOL Strike/Recce Aircraft based on P.1127. Pegasus 5 without PCB	December 1961
P.1157	Supersonic V/STOL Drone Aircraft. One RB.153 engine	January 1962
P.1158	Supersonic V/STOL Interceptor/Strike Aircraft. 2 x Pegasus 6 with PCB (Based on P.1155)	March 1962
P.1159	G-91 Replacement. (German Air Force requirement) with BS.94/4 and 2 x RB.162	March 1962
P.1160	Supersonic V/STOL OR.345 type aircraft. One Bristol Olympus 22R	March 1962
P.1161	P.1161 – [Not used. Focke-Wulf project number]	
P.1162	G.91 Replacement. (German Air Force Requirement) with Pegasus 5	March 1962
P.1163	G.91 Replacement using RB.168 development with PCB. (Became HS1170 in 1963)	March 1963
P.1164	P.1164 – P.1169 – Not used due to change of numbering with creation of HSA in 1963	
	Numbering nomenclature changed here with creation of Hawker Siddeley Aviation within the Hawker Siddeley Group	
HS.1170	V/STOL Strike/Recce Aircraft with BS.94/5 engine. (was P.1163)	September 1963
HS.1171	VG Trainer. Allocated to Hamble	

Project Number	Description	Date of Issue
HS.1172	Hunter. (Two Seat Hunter for P.1154 Nav/Attack System development)	
HS.1173	Advanced Trainer to AST.362 with single RB.172 -57 AR or 2 x RB.172/T260 engines	1965
HS.1174	Dual Version of P.1127 (RAF) to ASR.386. Pegasus 6. (Harrier T.2)	September 1965
HS.1175	2nd Generation Subsonic V/STOL Strike Aircraft. One Pegasus 6 development plus one lift engine (RB.162 or 189)	January 1967
HS.1176	Subsonic V/STOL Strike Aircraft. Developed Harrier for ASL presentation in USA	March 1967
HS.1177	Supersonic V/STOL Strike/Reconnaissance Aircraft. Pegasus with reheat plus Rolls/Allison XJ.99 or twin R-R Spey option	July 1967
HS.1178	Subsonic V/STOL Strike Aircraft. Pegasus, with PCB for VTOL only.	March 1968
HS.1179	Supersonic V/STOL Strike Fighter. Various engines. R-R RB.199 or 2 x RB.199, both with PCB and reheat or R-R Pegasus 15 with PCB. Plus VG studies. Kingston/Brough	April 1968
HS.1180	Allocated by Mr. Laight to the Spey Engine version of P.1154	
HS.1181	Wide-Speed range Aircraft with 'pop-out' lift fans. Pegasus 9D plus 2 x RB.202-10	October 1968
HS.1182	Basic Trainer. One Adour (no reheat). (Hawk) Flew 8/74	October 1968
HS.1183	Strike/Trainer. Two Vipers (no reheat)	September 1969
HS.1184	Advanced Harrier Development/Super Harrier (AV-16S). Pegasus 15	December 1969
HS.1185	Supersonic V/STOL Strike/Recce Aircraft (AV-16S). Pegasus 15 with PCB	December 1969
HS.1186	Advanced Harrier. Pegasus 16C	August 1970
HS.1187	Supersonic V/STOL Aircraft. Novel planform options inc. twin boom. Pegasus 16C with PCB and/or reheat	September 1970
HS.1188	Advanced Harrier to HIPAAS (High Performance Advanced Attack system). Pegasus 16. MDC/Kingston	July 1970
HS.1189	Simple Strike Aircraft. RB.199. Kingston/Brough	May 1971
HS.1190	Supersonic STOL Aircraft. RB.199R. Kingston/Brough	September 1971
HS.1191	Base Burning with and without lift engines. RB.199R plus XJ.99 type. Kingston/Brough	October 1971
HS.1192	Supersonic V/STOL Aircraft. RB.231 deflected turbojet plus reheat plus RB.227 lift engines. Kingston/Brough	January 1972

Project Number	Description	Date of Issue
HS.1193	Supersonic V/STOL Ogival Delta Wing Aircraft with canard. Pegasus 15 with PCB	September 1972
HS.1194	Super STOL Aircraft with canard. RB.199 with rotary augmentor	October 1972
HS.1195	STOL Strike Aircraft to AST.396 phase 2. Spey Development	August 1973
HS.1196	MINICAS. Various engines including Viper and RB.401	August 1973
HS.1197	Allocated to Brough for Buccaneer development	
HS.1198	Variable Cycle VT Aircraft. RB.406	December 1974
HS.1199	Executive 5 seat Harrier. Pegasus 11	June 1975
HS.1200	Air Superiority Fighter. Various engines including RB.199, RB.238, RB.246 etc.	July 1975
HS.1201	Air Superiority Fighter with variable incidence wing to AST.403. 1 x RB.199	
HS.1202	Air Superiority Fighter. 2 x RB.199 or 1 x RB.471	
HS.1203	Air Superiority Fighter. RB.231	November 1975
HS.1204	Allocated to Brough	
HS.1205	Air Superiority Fighter. Inc option with forward sweep and canard. Pegasus with PCB	June 1976
HS.1206	Allocated to Brough	
HS.1207	Allocated to Brough	
	Numbering nomenclature changed here with nationalisation of Hawker Siddeley Aviation and BAC to create British Aerospace	
P.1208	Subsonic V/STOL. Pegasus variant	August 1978
P.1209	PCB Demonstrator/Testbed. PCB Pegasus	August 1978
P.1210	Allocated to Brough	
P.1211	Allocated to Brough	
P.1212	Air Superiority Fighter with twin boom fuselage. PCB Pegasus 11	July 1979
P.1213	V/STOL Canard with Swept Back wings. Pegasus	
P.1214	V/STOL Canard with Swept Forward Wings. Pegasus	
P.1215	Allocated to Brough	
P.1216	ASTOVL Project (P.1212 with tail). Kingston/Brough	July 1980
P.1217	STOVL Aircraft with RALS PW.78-01	December 1979
P.1218	USN F-14/A-6 Replacement (model 279 via MDC)	August 1981
P.1219	Lightweight ASTOVL Aircraft	December 1981

Project Number	Description	Date of Issue
P.1220	Canard Lightweight V/STOL Aircraft	January 1982
P.1221	Half Scale P.1216	December 1982
P.1222	Tandem Fan VSTOL Fighter with canard. (R-R Bristol engine)	April 1983
P.1223	Augmented Thrust STOVL	May 1983
P.1224	Unmanned Fighter (Ferret). Viper or Adour 811 engine	May 1983
P.1225	Single Seat Hawk Combat Aircraft for SAAD 25	November 1983
P.1226	Subsonic STOVL design based on P.1216 layout	February 1984
P.1227	Development of Harrier GR.5 for NST.6464	January 1985
P.1228	Supersonic PCB Engine Canard for NST.6464	August 1985
P.1229	Supersonic Dry Engine P.1184 type for NST.6464. RB.533 engine	August 1985
P.1230	Supersonic PCB engine P.1205 type for NST.6464. RB.422	August 1985
P.1231	Supersonic Harrier (AV-16S) revived	September 1985
P.1232	Sea Harrier with GR.5 Rear Fuselage	September 1985
P.1233	SABA Prop Fan Anti-Helicopter Aircraft. Avco Lycoming T-55	May 1986
P.1234	SABA Jet Anti-Helicopter Aircraft. R-R Adour RT-172	May 1986
P.1235	Spey engined Fixed Wing P.1201-9 (AX proposal)	August 1986
P.1236	SABA, Small Agile Battlefield Aircraft	August 1986
P.1237	RALS with tailplane	September 1986
P.1238	Harry Fraser-Mitchell's Prop-Fan AHA/SABA	January 1987
P.1239	Versatile SABA	June 1987
P.1240	RESA, Reverse Flow Stealth ASTOVL	July 1987
P.1241	RALSA, Remote Augmented Lift Stealth ASTOVL, (RALS)	August 1987
P.1242	STOL Air Superiority Design, (ASTOVL comparison). RB.599-02	June 1988
P.1243	UFA	September 1988
P.1244	Harrier IIIA, P.1227 development with new radar, wing etc and can carry MSOW	May 1989
P.1245	Harrier III B&C, P.1231 with dry engine, new radar etc.	May 1989
P.1246	As P.1245 with PCB Pegasus	

Notes

Chapter 1
1. Mason, Francis K., *Hawker Aircraft Since 1920* (Putnam, 1991), p.1.
2. Davis, Mick, *Sopwith Aircraft*, (Crowood Press, 1999), p.7.
3. *Hawker Association Newsletter*, No.12, 2006.
4. Davis, p.8; Mason, p.4.
5. Numbers built vary between sources as there is confusion due to contracted numbers differing from those actually built. Sub-contract was widespread as was licence production among the Allied nations, making accuracy challenging. Machines were also frequently transferred between RFC and RNAS, making double counting a further area of confusion.
6. Davis, p.11; Mason, p.2.
7. Mason, p.6.
8. Mason, pp.8-9; Sweetman, John, *Sydney Camm: Hurricane and Harrier Designer* (Pen & Sword, 2019), pp.15-16.
9. Fozard, Dr John W., *Sydney Camm and the Hurricane* (Airlife, 1991), pp.59 & 133; Tuffen, Harold J. & Tagg, Albert E., *The Hawker Hurricane: Design, Development and Production* (Royal Aeronautical Society, 1985), p.13.
10. Mason, pp.55-6.
11. Manhours for Hurricane construction 10,000; Spitfire 15,000. Quoted in Sweetman, *Sydney Camm – Hurricane and Harrier Designer*, p.87.
12. Fozard, p.138; Fozard (ed), *Proceedings of the Hurricane 50th Anniversary Symposium* (Royal Aeronautical Society, 1985), pp.72-3.
13. Brochure for Hawker Aircraft Share options, 1933. HSA/MAN/008 BAE Systems, courtesy of Brooklands Museum.
14. Barnett, Correlli, *The Audit of War* (Papermac, 1987), p.130.
15. Budgen, Christopher, *Hawker's Secret Cold War Airfield: Dunsfold, Home of the Hunter and Harrier* (Pen & Sword, 2020), p.6.
16. Barnett, pp.210-11.

Chapter 2
1. Turnhill, Reginald and Reed, Arthur, *Farnborough: The Story of RAE* (Hale, 1980), p.41.
2. Mason, op. cit., p.6.
3. Hawker legal documents HAL/HIS/272-5. BAE Systems, courtesy of Brooklands Museum.
4. Farara, Chris, Vivian Stanbury: draft obituary, Biographical Notes, HAL/HIS/007, Brooklands Museum.
5. *Hawker Association Newsletter*, No. 13, 2006.
6. Hooper, Ralph, Robert Marsh obituary, in *Aerospace*, the Royal Aeronautical Association publication, February 1996, 'Flight Testing the Early Jets'; Robert Marsh, HAL/HIS/162. Brooklands Museum.

7. Hooper, Ralph, 'Camm's Contribution and Legacy, in *Sydney Camm and the Hurricane*, edited by John Fozard (Airlife, 1991), p.202; John Fozard obituary, Sir Robert Lickley. *Aerospace*, the Royal Aeronautical Association publication, September 1996.
8. Mason, p.vii, op. cit., p.204.
9. Hooper, 'Camm's Contribution and Legacy,' op. cit., p.207.
10. Braybrook, Roy, personal communication, March 2021.
11. Buss, Brian, personal communication, October 2020.
12. Hooper, Ralph, National Life Stories/British Library, C1379/27, p.180.
13. Hassard, David, *Making Them Right – A Brief Record of Charles Plantin's 40 Year Career at Hawker Aircraft, Kingston* (unpublished monograph, 2007), p.14.
14. Flint, Colin, 'The Mithreum Test Frame,' *Hawker Association Newsletter*, No. 33, 2002.
15. Hooper, Ralph, National Life Stories/British Library, C1379/27, p.139.
16. Braybrook, Roy, personal communication, March 2021.
17. Braybrook, Roy, *Harrier and Sea Harrier* (Osprey, 1984), p.42.
18. Design Manning Papers, HSA/MAN/001, BAE Systems, courtesy of Brooklands Museum.

Chapter 3
1. *Science at War*, HMSO, 1945, p.1.
2. Camm Diaries, HAL/HUN/063, BAE Systems, courtesy of Brooklands Museum.
3. David Lockspeiser Papers, Brooklands Museum.
4. Fozard, 'Sydney Camm and the Hurricane,' op. cit., p.204.
5. National Archives, Kew, AIR2/13058; Appendix C, HAL/HUN/062, BAE Systems, courtesy of Brooklands Museum.
6. Ibid.
7. Ibid.
8. Camm Diaries, HAL/HIS/082, BAE Systems, courtesy of Brooklands Museum.
9. Frank Mason dates this as 1955 due to a typo on the project list. The project number identifies it as a 1953 design as corrected later in Tuffen's master list.

Chapter 4
1. P.1093 GA. P.1097 GA. BAE Systems, courtesy of Brooklands Museum.
2. Camm Diaries, HAL/HIS/083. BAE Systems, courtesy of Brooklands Museum.
3. P.1103-8 GAs, BAE Systems, courtesy of Brooklands Museum.
4. P.1103 various GAs, BAE Systems, courtesy of Brooklands Museum.
5. HAL/PRJ/018, BAE systems, courtesy of Brooklands Museum.
6. National Archives (NA), Kew, AIR20/9042.
7. Ibid.
8. Camm Diaries, HAL/HIS/083, BAE Systems, courtesy of Brooklands Museum.
9. HAL/PRJ/057, BAE Systems, courtesy of Brooklands Museum.
10. NA, AIR20/9042.
11. Camm Diaries, HAL/HIS/083, BAE Systems, courtesy of Brooklands Museum.
12. Camm Diaries, HAL/HIS/083, BAE Systems, courtesy of Brooklands Museum.
13. NA, AIR20/9042.
14. Ibid.
15. Ibid.
16. Ibid.
17. Ibid.
18. Ibid.

19. GOR.339 – TSR/RQM/001, Brooklands Museum.
20. HAL/HIS/083, BAE Systems, courtesy of Brooklands Museum.
21. Buss, Brian, personal communication, November 2020; *Hawker Association Newsletter*, No.54, 2019.
22. Camm Diaries, HAL/HIS/083, BAE Systems, courtesy of Brooklands Museum.
23. Ibid.
24. Simpson Report, HAL/PRJ/077, BAE Systems, courtesy of Brooklands Museum.
25. HAL/HIS/083, BAE Systems, courtesy of Brooklands Museum.
26. Williams, Dr Geoffrey et al, Crisis in Procurement: A Case Study of TSR-2 (Royal United Service Institution, 1969), p.18; HAL/PRJ/055, Brooklands Museum.
27. Ibid., p.18; HAL/PRJ/052, BAE Systems, courtesy of Brooklands Museum.

Chapter 5
1. Ashwood, P.F., 'Powerplant Requirements for Vertical Take-Off Aeroplanes Using Jet Lift', NGTE, VTL/MIS/001, Brooklands Museum.
2. Camm Diaries, HAL/HIS/083, BAE Systems, courtesy of Brooklands Museum.
3. Hooper notes, HAL/HIS/027, Brooklands Museum.
4. Camm Diaries, HAL/HIS/083, BAE Systems, courtesy of Brooklands Museum.
5. NBMR-3 requirement, HAL/PRJ/020, BAE Systems, courtesy of Brooklands Museum.
6. Hooper Notes, HAL/PRJ/020-2, Brooklands Museum.
7. Hooper Notes, HAL/HIS/027, Brooklands Museum.
8. Ibid.
9. Miscellaneous Operational Requirements folder, Brooklands Museum.
10. Hawker P.1154 Design Study Submission to Joint Naval/Air Staff Requirement AW.406/OR.356, HAL/PRJ/028, BAE Systems, courtesy of Brooklands Museum.
11. The Influence of Layout on the Aerodynamic Characteristics of European Jet Lift Aircraft, R.S. Williams, May 1984. BAe/VTL/026 BAE Systems, courtesy of Brooklands Museum.
12. NA, Kew, AIR19/1053.
13. Camm Diaries, HAL/HIS/084, BAE Systems, courtesy of Brooklands Museum.
14. NA, Kew, AIR19/1053, op. cit.
15. Ibid.
16. It should be noted that, during the design process, many different drawings were produced giving slightly different dimensions etc. Only some of these are reproduced here.
17. NA, Kew, AIR19/1053.
18. Camm Diaries, HAL/HIS/084, BAE Systems, courtesy of Brooklands Museum.
19. Ibid.
20. Ibid.
21. Ibid.
22. Dow, Andrew, *Pegasus: The Heart of the Harrier* (Pen & Sword, 2009), p.278; HAL/PRJ/009, BAE Systems, courtesy of Brooklands Museum.

Chapter 6
1. SABA Project Papers, BAE/PRJ/201, BAE Systems, courtesy of Brooklands Museum.
2. Ibid.
3. Jones, Andy, personal communication, March 2022.

4. SABA Project Papers, BAE/PRJ/201, BAE Systems, courtesy of Brooklands Museum.
5. Jones, Andy personal communication, March 2022.
6. SABA Project Papers, BAE/PRJ/201, BAE Systems, courtesy of Brooklands Museum.
7. SABA Papers, New Business Paper No.2, HSA/PRJ/007, BAE Systems, courtesy of Brooklands Museum.
8. Preliminary specification BAE/PRJ/054, BAE Systems, courtesy of Brooklands Museum.
9. Jones, Andy, personal communication, March 2022.
10. *Air Force Magazine*, April 1988, BAE/PRJ/053, Brooklands Museum.
11. SABA Papers, Correspondence, BAE/PRJ/053, BAE Systems, courtesy of Brooklands Museum.
12. Ibid.
13. Ibid.
14. Ibid.
15. Ibid.
16. Ibid.
17. Hooper notes, BAE/PRJ/017, Brooklands Museum.

Chapter 7
1. Dow, op. cit., pp.380-3.
2. HSA/PRJ/043, BAE/PRJ/003, BAE Systems, courtesy of Brooklands Museum.
3. Ibid., pp.432-3.
4. Inchbold, Guy, 'Outside Edge,' *The Aviation Historian*, No. 38, 2022, p.106.
5. BAe Project Papers, BAE/PRJ/006, BAE Systems, courtesy of Brooklands Museum.
6. Ibid.
7. BAe Project Papers – Political, BAE/PRJ/008, BAE Systems, courtesy of Brooklands Museum.
8. Hooper notes, BAE/PRJ/007, Brooklands Museum.
9. Ibid.
10. BAe Project Papers, BAE/PRJ/005, BAE Systems, courtesy of Brooklands Museum.
11. HAL/HIS/007, Brooklands Museum.

Chapter 8
1. Camm Diaries, HAL/HIS/083, BAE Systems, courtesy of Brooklands Museum.
2. Project Programme 1973, HSA/PRJ/005, BAE Systems, courtesy of Brooklands Museum.
3. UFA Papers, 1983, BAE/PRJ/185, BAE Systems, courtesy of Brooklands Museum.
4. Rich, Ben R., and Janos, Leo, *Skunk Works* (Warner, 1994), pp.253-4.
5. UFA Papers, 1983, BAE/PRJ/185, BAE Systems, courtesy of Brooklands Museum.
6. Ibid.
7. Ibid.
8. UFA Design Papers BAE/PRJ/189, BAE Systems, courtesy of Brooklands Museum.
9. Ibid.
10. 'Air Combat Effectiveness of P.1224', BAE/PRJ/186, BAE Systems, courtesy of Brooklands Museum.
11. Ibid.
12. Hooper Notes, BAE/PRJ/017, Brooklands Museum.

Index

A&AEE Boscombe Down, 32, 56
Advanced Projects Group (HSA), 103

Aircraft
British Aerospace;
  AV-8B (MDC), 155, 164, 191
  Hawk, 39–40, 44–5, 130–1, 136, 141, 143, 146–9, 155, 201, 209, 211
  Harrier GR.5, 164, 166, 169, 191, 211
  Harrier GR.7, 168
  Jaguar, 130, 153, 156, 166, 169
  Sea Harrier, 43, 127, 166, 168, 211, 214
  T-45 (MDC), 44, 191

Hawker;
  Audax, 18
  Duiker, 9, 24
  Fury biplane, 13
  Fury monoplane, 16, 202–203
  Hardy, 18
  Hart, 13, 57
  Henley, 18, 202
  Heron, 13
  Hind, 131, 199
  Hornbill, 13
  Hunter, 36, 39–41, 45, 47–52, 54, 56–72, 74–6, 104, 110, 112, 115, 117–18, 129–30, 173–4, 195–8, 201, 204–206, 209, 213
  Hurricane, 13–16, 18, 27, 38, 41, 57, 130, 192, 202, 213–14
  Sea Fury, 16, 19
  Sea Hawk, 16, 60, 201
  Tempest, 16, 41, 189, 202–203
  Tornado, 15, 153, 169
  Typhoon, 15, 18, 38, 41, 130, 170, 188, 202
  Woodcock, 9, 24

Hawker Siddeley;
  Harrier, 33, 39–40, 43, 100, 102, 106, 113, 118, 126–8, 130, 143, 151–6, 159, 161–4, 166, 168–9, 195, 201, 209, 211, 213–15
  Kestrel, 43, 112, 122, 126
  P.1127, 33, 37, 39, 41, 43, 45, 64, 102, 104–107, 111, 118, 126, 151, 206, 208–209

Sopwith;
  1½ Strutter, 5
  Camel, 6
  D1, 5
  Dolphin, 6
  Pup, 5
  Snail, 15
  Snark, 15
  Snipe, 6, 9
  Sparrow, 80, 172
  Tabloid, 5
  Triplane, 6

Other;
  Aérospatiale/BAC Concorde, 196
  Avro 730, 74, 88, 94
  Avro Vulcan, 64, 120
  BAC TSR.2, 21, 49, 97–9, 120, 125, 127, 193, 199
  Boeing AH-64 Apache, 143
  Boeing B-17, 172
  Boeing B-29, 198
  Boeing E3 AWACS, 176, 180
  Bell Boeing V-22 Osprey, 146
  Bell XV-15, 143
  Convair F-102 Delta Dagger, 91
  Convair F-106 Delta Dart, 91
  Dassault/Dornier Alpha Jet, 131
  Dassault Mirage IIIV, 109
  de Havilland DH.82 Tiger Moth, 172
  de Havilland DH.82B Queen Bee, 172
  de Havilland Vixen/Sea Vixen, 49, 106, 110, 118
  EFA, 143, 170–1, 188
  English Electric Canberra, 32, 83, 85–6, 88
  English Electric Lightning, 32, 48–9, 56–7, 61, 100
  Gloster Javelin, 49, 64
  Grumman F-14 Tomcat, 176, 210
  Lockheed F-117A Nighthawk, 183
  Lockheed SR-71 Blackbird, 172
  Lockheed Martin F-35A Lightning II, 168
  Lockheed Martin F-35B Lightning II, 100, 151, 168
    McDonnell Douglas F-15 Eagle, 176

McDonnell Phantom, 102, 119–21, 124, 127, 199
Mikoyan-Gurevich MiG-15, 57–8
Myasishchev M-4 Bison, 59
Pilatus PC9D, 143
Republic A-10 Thunderbolt II, 129
Republic F-105 Thunderchief, 99, 196
F-111, 125, 177
Ryan X-13 Vertijet, 179
Short SC.1, 101
Sukhoi Su-25 Frogfoot, 150
Sukhoi Su-27 Flanker, 183
Supermarine Spitfire, 15, 41, 213
Supermarine Swift, 61
Tupolev Tu-4, 198
Tupolev Tu-95 Bear, 59

Aircraft Manufacturers
Armstrong Whitworth, 19, 24, 44, 76, 80
Avro, 19, 37, 44, 64, 74, 76, 88, 94, 96, 98
BAC, 22, 32, 97–8, 109, 153, 210
BAE Systems, 46, 151, 189, 213–16
Bell, 73, 87
Blackburn, 21, 32, 44, 106
Boeing, 198
Breguet, 107
Bristol Aeroplane Co, 6, 74, 80
British Aerospace (BAe), 1, 18, 21–2, 32–3, 35, 44–6, 130, 145, 149–51, 157, 165–6, 168–70, 177, 184, 189, 191, 200, 202, 210, 215–16
Convair, 91
Dassault, 109
de Havilland, 21, 44, 48–9, 63, 75–6, 79, 81, 83, 93, 95, 106, 172
English Electric, 22, 32, 49, 56–7, 61, 76, 80, 97, 99
Fairey, 27, 31, 48, 76, 80, 92
Focke-Wulf, 107, 208
Fokker-Republic, 107
Folland, 21, 44
General Dynamics, 177
Gloster, 9, 18, 19, 24, 32, 44, 49, 64
Handasyde, 9, 23
Hawker Aircraft Ltd, 1, 6, 11, 14, 17–19, 26, 31, 40, 49, 57, 67, 102, 197
Hawker Siddeley Aviation, 8, 17–19, 21–2, 31, 43–4, 46, 64, 76, 81, 94–8, 100, 102–103, 107, 109, 116, 118–19, 121–2, 124, 128, 151, 155, 173, 194, 201–202, 208, 210
HG Hawker Engineering Co, 1, 3, 7, 11, 18, 23, 25, 28–9
Howard Wright, 2–3
Lockheed, 141, 172, 177, 183

MacAir, 168
Martinsyde, 9
McDonnell, 102, 121, 153, 155
McDonnell Douglas, 153, 155
Miles, 73
Myasishchev, 59
Panavia, 153, 169
Saunders Roe (Saro), 76
Shorts, 95, 99, 101
Sopwith Aviation Co, 2–9, 11, 15, 17–20, 23–4, 34, 172, 213
Tupolev, 59
Vickers Supermarine, 76, 97, 99

Air Ministry, 11, 42–3, 46, 49, 52, 55, 57–60, 67, 70, 74, 77, 80–1, 83, 85–6, 95–8, 102, 114, 117, 122, 197
ASTOVL, 127, 150–1, 156, 166, 170, 197, 210
Bawdsey, 47
Brooklands, 2–3, 5–6, 9, 14, 17, 19, 23, 213–16
Brough, 22, 32–3, 43–4, 138, 153, 209–10
Canbury Park Road, 4, 6, 9, 11, 16–17, 20, 24, 27, 29, 31
China, 145
combat air patrol, 113
Cranwell, 96
CTOVL, 165
CVA-01, 127, 199
DoD, 169

Drones/UCAV
GAF Jindivik, 172–3
Lockheed D-21, 172
Tempest, 189
UAV, 172, 189
UCAV, 172

Dunsfold, 18–20, 22, 42, 44, 49, 56, 60–1, 102, 104, 122, 131, 150, 189, 191, 200, 213
EKCO, 48
ESM, 180

Engines
Allison T-56, 138
Allison XJ-99-RA-1, 151

Armstrong Siddeley:
Sapphire, 57–8, 64–5, 67, 195, 204–206
Snarler, 58, 65, 204–205
Viper, 178, 181–2, 210–11
Avco Lycoming ALF-502, 135, 142
Avco Lycoming T-55, 132, 134, 138, 211

Blackcap rocket motor, 178
Bristol:
   BE.53, 99, 102, 206–207
   BS.53, 102, 105, 151, 208
   BS.53/6, 105
   BS.100, 106–108, 112, 115, 119–23, 126–7, 164, 208
   BS.100/3, 106
   BS.100/8, 106, 112, 115, 119, 121–2, 208
   BS.100/9, 106, 108, 112, 120, 208
   Olympus, 86, 94–5, 120, 208
   Orpheus, 69, 101, 206
   Pegasus, 37, 104, 106, 126, 151–7, 159, 162, 164, 168, 195–6, 208–11, 215

de Havilland:
   Ghost, 79
   Goblin, 79
   Gyron, 63, 75–7, 79–82, 85, 91, 94–5, 195–6, 205–207

General Electric T-64, 138
General Electric TF-34, 138

Napier Double Scorpion rocket motor, 207

Orenda, 76

Rolls-Royce:
   Adour, 135, 137, 140, 144, 177, 181, 186, 209, 211
   Avon, 56–7, 60–1, 64–8, 70–1, 91, 195–6, 204–207
   Conway, 86, 88
   Merlin, 14
   RA.28, 65
   RB.106, 64, 74, 205
   RB.108, 99, 101
   RB.112, 75, 77, 205
   RB.133, 91, 206
   RB.146, 68, 207
   RB.153, 103, 174, 207–208
   RB.163, 103, 207
   RB.168-31D, 152
   RB-199, 152
   RB.401-06, 181
   RB.422-48, 158
   RB.422-60, 162, 164, 195
   Spey, 121

Engine Manufacturers
   Allison, 138, 151, 209
   Armstrong Siddeley, 19, 59, 76, 204
   Bristol Siddeley Engines (BSEL), 43, 69, 86, 94–6, 99, 101–103, 105–108, 120–3, 126, 196, 206–208, 211
   Napier, 65, 102, 206, 207
   Rolls Royce, 2, 14, 31, 57, 59–60, 63–6, 74–6, 85, 88, 91, 96, 99–101, 103, 119–21, 124, 130, 135, 138, 151–2, 154, 158–9, 164, 166, 174, 177, 181, 189, 195

Farnborough, 23–4, 32, 38–9, 41–2, 49, 59, 130, 171, 200, 213
Ferranti, 37, 122
Ferret, 175
Fleet Air Arm, 65
FLIR, 180, 183
Future Projects Office, 45, 134, 169, 171, 174

Guided Missiles
   AGM-12 Bullpup, 50, 121
   AIM-4 Falcon, 80
   AIM-9 Sidewinder, 49
   AIM-54 Phoenix, 176
   AIM-120 AMRAAM, 161, 198
   ASRAAM, 138, 147, 161
   BAe Rapier, 175
   BAe Sea Eagle, 161
   Blue Jay, 48–50, 61, 65, 76, 198, 206
   Blue Sky, 48
   Blue Vesta, 48, 76, 198
   Bristol Bloodhound, 122
   de Havilland Firestreak, 48, 50, 52, 54, 58, 60, 65, 67–8, 207
   Fairey Fireflash, 48, 58–9, 68
   HS Dynamics/Matra AS.37 Martel, 121
   HS Dynamics Red Top, 48, 76, 113, 117, 122, 199
   LRAAM, 197
   MRAAM, 176
   Red Hebe, 76, 80, 198
   SAM, 176, 195

Ham, 6–7, 20–1
Hamble, 44–5, 121–2, 124–5, 208
Hughes, 122, 198

India, 29, 68, 145, 194, 207
IRST, 187

Janes, 138, 145

Kingston upon Thames, 1, 4, 6–7, 9, 11, 16–17, 20–2, 24, 27–8, 32–3, 36–7, 38–46, 57, 59, 60, 64, 75, 80–1, 84, 88, 93–4, 98, 100, 103, 105–107, 110, 112, 114, 116–19, 121–2, 124, 130, 137–8, 147, 150–1, 153, 155, 157, 161–2, 164–6, 168–71, 174, 183, 188–9, 191–2, 200, 202, 209–10, 214

Langley, 15, 17, 19, 24, 32, 38, 42
Leavesden, 94
Leonado, 189
LERX, 156, 158–9
LWR, 187

Malvern, 77, 91
Marine Corps US, 154
Mayford, 9
MBDA UK, 189
Minicas, 130
MSOW, 143, 211

NASA, 169
NATO, 59, 73, 100, 107, 109–10, 129–30, 152, 198–200
NGTE Pyestock, 100, 215

Olympia Show, 5

Operational Requirements/Specifications
   AST.396, 153, 156, 210
   AST.403, 156, 210
   AST.410, 166
   AST.412, 131, 133
   ER.134T, 64, 74
   F.119D, 60
   F.155T, 76–7, 79–81, 97
   GOR.339, 83, 85–6, 89–91, 93–9, 207, 215
   NBMR-3, 107, 109–10, 215
   NST.6464, 166, 169, 211
   OR.23, 94
   OR.24, 88
   OR.228, 54, 56
   OR.329, 75, 80, 84, 88
   OR.345, 104, 106, 110, 208
   OR.356, 106, 110, 112, 215
   TE.6/50, 79

People
   Allen, John, 130
   Atherton, 77
   Barnett, Correlli, 19–20, 213
   Barrett, Joe, 27, 66, 68
   Bennett, Frederick Ibbotson, 1, 3
   Bowen, Dr E.G., 47
   Braybrook, Roy, 33–4, 43–4, 102–103, 214
   Brown, Eric, 110
   Brown, George, 124
   Buss, Brian, 35, 91–3, 214–15
   Camm, Sir Sydney, 9, 13–16, 20–1, 24–32, 34, 36–8, 40–4, 55, 80–1, 83–4, 88, 91–7, 101, 103, 110–11, 124, 192–3, 200, 213–16
   Carter, George, 9, 24
   Crampton, John, 2

Davies, Handel, 81
Dobson, Roy, 96
Douglas-Home, Sir Alec, 122
Eden, Sir Anthony, 86
Emson, Sir Reginald, 118
Eyre, William, 1, 3
Farara, Christopher, 131, 213
Farley, John, 138
Fozard, John, 20, 32, 35, 37, 43–4, 58, 64–5, 69–70, 74–5, 77–9, 81, 103, 106, 111, 126, 173, 193, 213–14
Fraser-Mitchell, Harry, 133–4, 211
Fredericksen, DN, 141
Gillibrand, Sydney, 130
Hall, Sir Arnold, 124, 130–1
Hancock, Victor, 134–5, 155–6, 159, 177, 179, 181
Hansford, Christopher, 45, 99, 134, 165, 171, 174–6, 180, 182
Hartley, AVM, 124
Hawker, Harry, 1–2, 5, 7–9, 23
Healey, Denis, 124
Hooker, Sir Stanley, 101, 106, 126
Hooper, Ralph, 20, 32–4, 37, 43–4, 67, 75, 77, 79, 81, 83, 96, 102–106, 110–11, 126, 128, 131, 133, 149, 165, 168–9, 189, 193, 197, 213–16
Huckle, Miles, 143, 156
Hudson, Gordon, 44, 134
Jackson, Wg Cdr, 86
Jenkins, Roy, 122, 124
Johnson, Gp Capt, 86, 131
Jones, Andy, 131–6, 138, 147–9, 215–16
Jones, E.T., 81
Kirkpatrick, Herbert James, 83–4, 87–8
Kyle, Wallace, 61
Laight, Barry, 111, 209
Liddell, P.W., 147
Locke, King Hugh, 2
Lockspeiser, Ben, 73
Lockspeiser, David, 49, 214
Lowe, Wg Cdr, 88
Lucas, Philip, 15
Lygo, Sir Raymond, 138, 141
Macmillan, Harold, 86
Mansell, Mick, 148
Marsh, Robert, 32, 42, 91, 103, 213
Martin, H.P., 9, 23
Mason, Francis, 34, 102, 213–14
McGuire, Air Commodore, 86–7
Mills, J.D., 65, 68
Milsom, Len, 130, 148–9
Moolgavkar, Gp Capt, 68
Murphy, Frank, 68
Nicholson, 114
Pelley, Sir Claude, 88

Pike, Sir Thomas, 80–1
Rich, Ben, 177, 216
Rolls, C.S., 2
Roulston, Air Commodore, 88
Sandys, Duncan, 68, 72, 86, 91, 95, 97, 107, 195
Satterley, Air Marshal, 83
Sigrist, Frederick, 1–3, 8, 13, 23, 24
Silyn-Roberts, Air Commodore, 61
Simpson, Duncan, 95, 215
Smith, Herbert, 6, 8
Sopwith, Thomas Octave Murdoch, 1, 8
Spriggs, Sir Frank, 96
Thomson, Capt Bertram, 9, 24
Thorneycroft, Peter, 111, 118, 120–1, 127
Towell, Keith, 135, 137, 146
Trenchard, Sir Hugh, 2
Tuffen, Harold, 26, 28, 32, 36, 92, 213–14
Turner, Mike, 148–9, 168
Tuttle, Sir Geoffrey, 61, 88, 93
Wallis, Sir Barnes, 153
Watkinson, Harold, 111
Watson, Watt Sir Robert, 47
Wibault, Michel, 102
Willder, Wg Cdr, 90
Wilson, Harold, 122, 127
Williams, Ron, 28, 35, 43–4, 91, 96, 101–103, 106, 215
Wilson, Harold, 122, 124, 127
Woods, T., 179, 180
Woodward-Nutt, 61
Zuckerman, Sir Solly, 121

Projects - Aircraft
Coventry
AW.406, 112, 215
AW.681, 106, 124
HS.681, 194

Kingston
F.3/48, 56–9
F.43/46, 54, 56, 204
F.44/46, 54
HS.1184, 153–4, 209
HS.1184-16, 153
HS.1185, 154, 209
HS.1205, 155–6, 210
P.1054, 55, 204
P.1067, 48, 56–7, 61, 204–205
P.1069, 57, 204
P.1071, 57, 204
P.1083, 58–62, 70, 205
P.1084, 64, 66, 205
P.1092, 64, 205
P.1093, 64, 205, 214
P.1096, 64, 74, 205
P.1097, 64, 74, 205, 214
P.1099, 61, 205
P.1099A, 56
P.1100, 64, 205
P.1101, 66, 205
P.1102, 65, 205
P.1103, 75–81, 97, 195, 197, 199, 205–206, 214
P.1104, 75, 77, 205
P.1105, 65, 206
P.1107, 78–9, 206
P.1109A, 48
P.1114, 66, 206
P.1116, 81, 206
P.1121, 21, 34–7, 42, 45, 67–9, 79, 81–97, 99, 101–102, 195–9, 206
P.1124, 173, 174, 206
P.1125, 91, 96–7, 206
P.1126, 101, 206
P.1128, 69–70, 206
P.1129, 96–9, 199, 207
P.1130, 68, 207
P.1132, 102, 207
P.1133, 68
P.1135, 68, 207
P.1136, 103, 207
P.1138, 103, 207
P.1139, 103, 207
P.1140, 103, 207
P.1143, 103, 207
P.1150, 104–108, 208
P.1152, 106, 111, 208
P.1154, 43, 45, 106–108, 110–12, 114–22, 124–8, 151, 164, 194, 196–7, 199, 208–209, 215
P.1157, 174, 208
P.1179, 152
P.1179L, 152
P.1179P, 153
P.1179U, 153
P.1187, 156
P.1205, 154, 156–7, 211
P.1212, 156–9, 210
P.1213, 157, 210
P.1214, 157, 160, 210
P.1216, 128, 150, 158–9, 161–71, 195–7, 210
P.1224, 175–6, 178, 180–6, 188, 190, 211, 216
P.1230, 169, 211
P.1233, 134–8, 143, 145–7, 211
P.1233/5, 143, 147
P.1234, 135, 137, 140, 142, 144, 211
P.1234/2, 135, 137

P.1238, 131–2, 134, 137, 211
SABA, 129, 132, 134, 136, 144, 146, 148, 150, 199, 211, 215–16
UFA, 172, 175, 180, 182, 184, 186, 188, 200, 211, 216

Kingston/MDC St Louis
AV-16E, 153

Warton
P.17A, 99
P.103, 166, 168
P.109, 166
P.112, 165–6, 169
P.115, 166, 169
P.116, 166, 169

PCB, 106–108, 112, 122, 126–7, 152–3, 155–7, 159, 162, 164, 196, 208–11
Philco, 49

Radar
AI.18, 80
AI.20, 48–9, 65–6, 206
AI.23, 48–9, 54, 68, 206–207
Chain Home, 47
Doppler, 88–9, 94
Green Willow (AI.20), 48
Yellow Lemon (Doppler), 88–90

RAE Farnborough, 32, 38–9, 41, 56, 58, 75, 101, 130, 169, 171, 200, 213
RALS, 166, 169, 210
Richmond Road, 20, 39–40
Royal Aero Club, 2
Royal Air Force (RAF), 2, 11, 54, 89
Royal Flying Corps (RFC), 2, 5, 17, 24, 213
Royal Naval Air Service (RNAS), 5, 213

Royal Navy (RN), 4, 16, 24, 103, 106–107, 110–14, 119–21, 127, 166, 168, 172, 199
Royal Radar Establishment (RRE), 77

Schneider Trophy, 5
Shoeburyness, 164, 196
Singapore, 50
Skyhook, 187
Soviet Union, 51, 53, 57, 59, 131, 177, 195, 198–200
STOVL, 43, 100, 154, 158, 160–1, 165–6, 210
Sweden, 49–50
Switzerland, 50

tandem fan, 166, 169
Thrust Measuring Rig (TMR), 100

US Army, 129
United States Air Force (USAF), 129, 145, 177
United States Marine Corps (USMC), 153–4

Violet Picture (UHF homer), 94
V/STOL, 34, 43, 45, 100, 102–108, 111–12, 115, 121, 126–7, 151–3, 155, 157, 168–9, 173, 195, 206–10
VTOL, 37, 64, 99–101, 103, 106–107, 120, 124, 127, 179, 206, 209

Warsaw Pact, 100, 129, 175, 177, 183–5, 187, 198
Warton, 22, 46, 138, 143, 148, 150, 162, 165–6, 168–71, 184, 188
Weybridge, 2, 9, 22, 35, 45–6, 133, 135, 137, 143, 153, 162, 171, 191

Yugoslavia, 145